Date _____

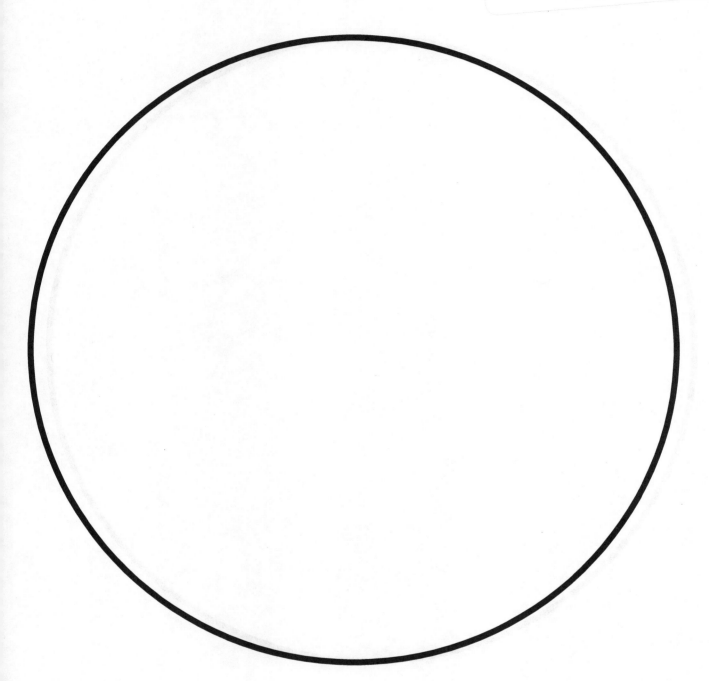

What is under my microscope?

_____

Date _____

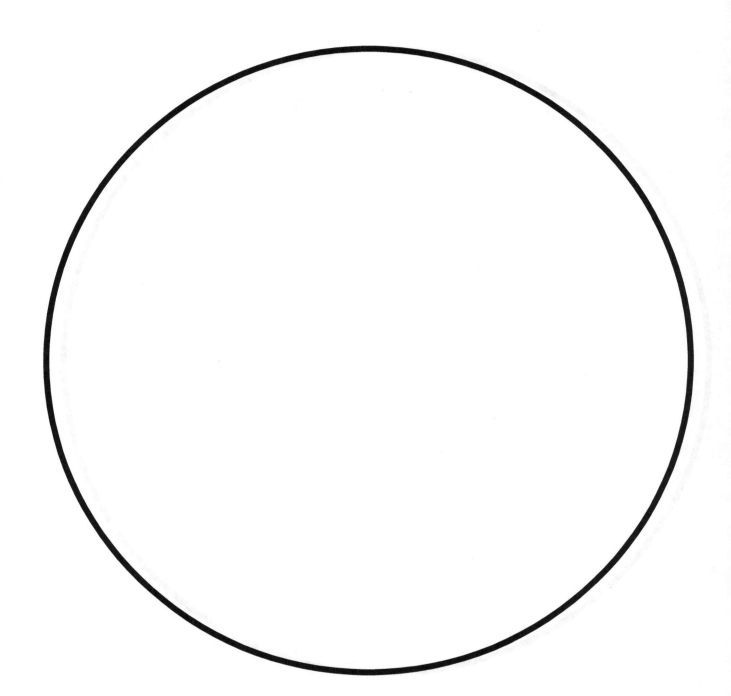

What is under my microscope?

_____

Date _____

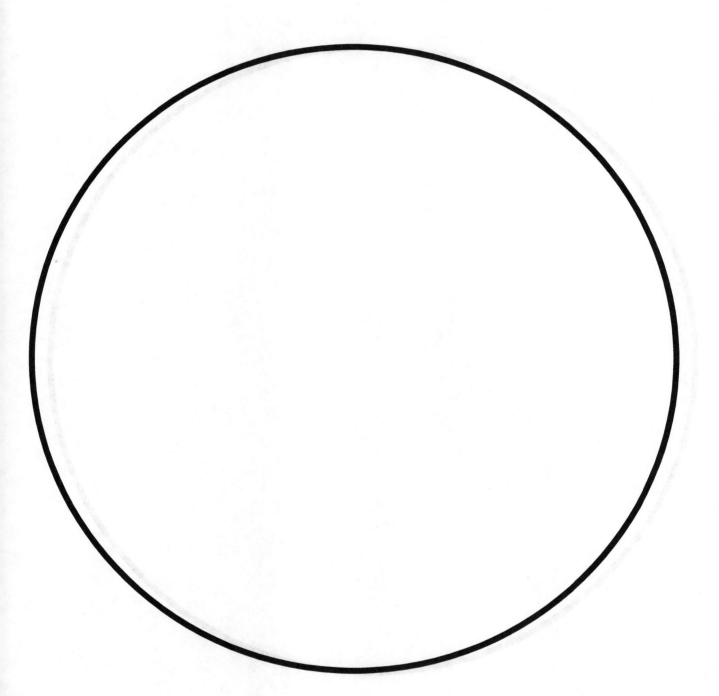

What is under my microscope?

_____

Date _____

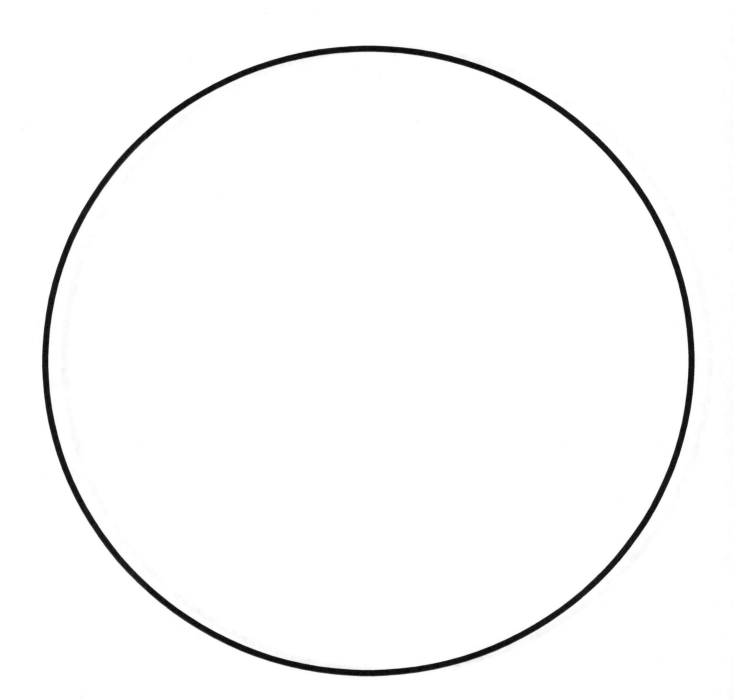

What is under my microscope?

_____

Date _____

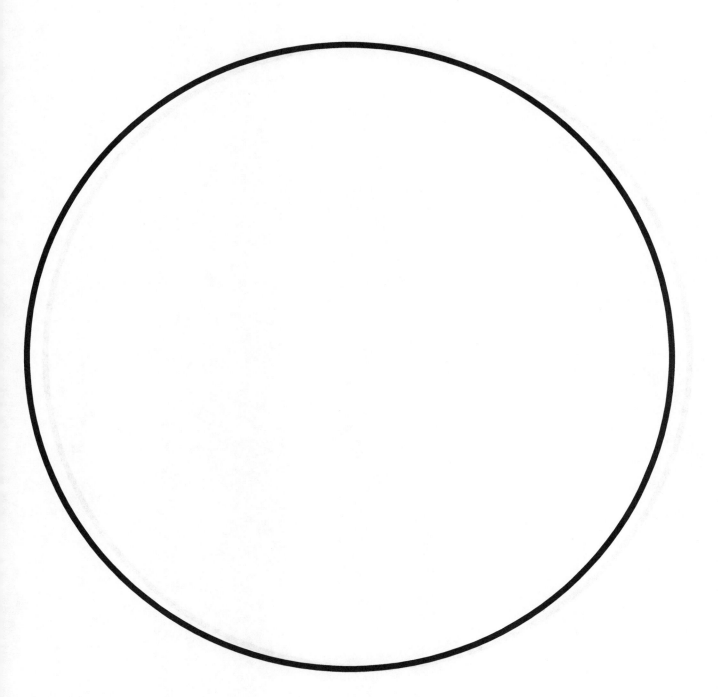

What is under my microscope?

_____

Date _____

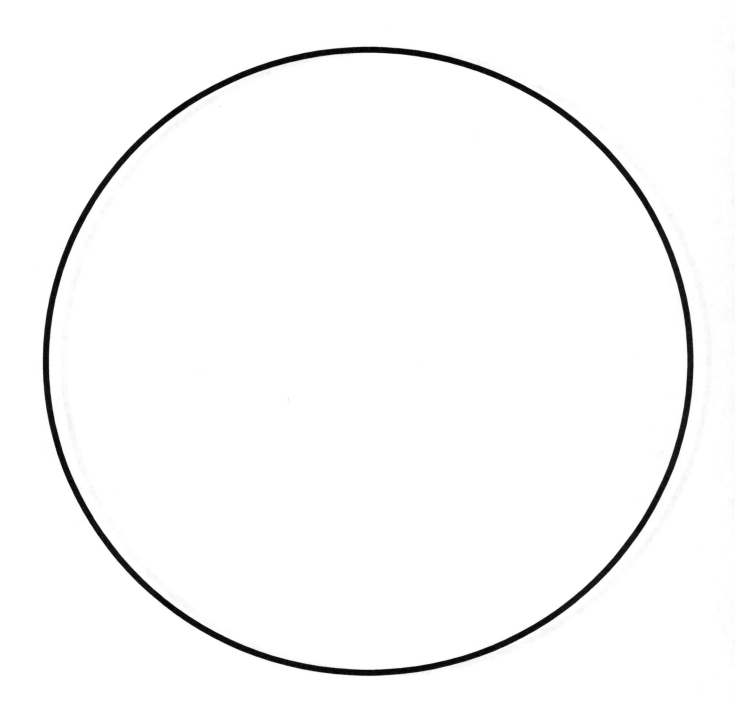

What is under my microscope?

_____

Date _____

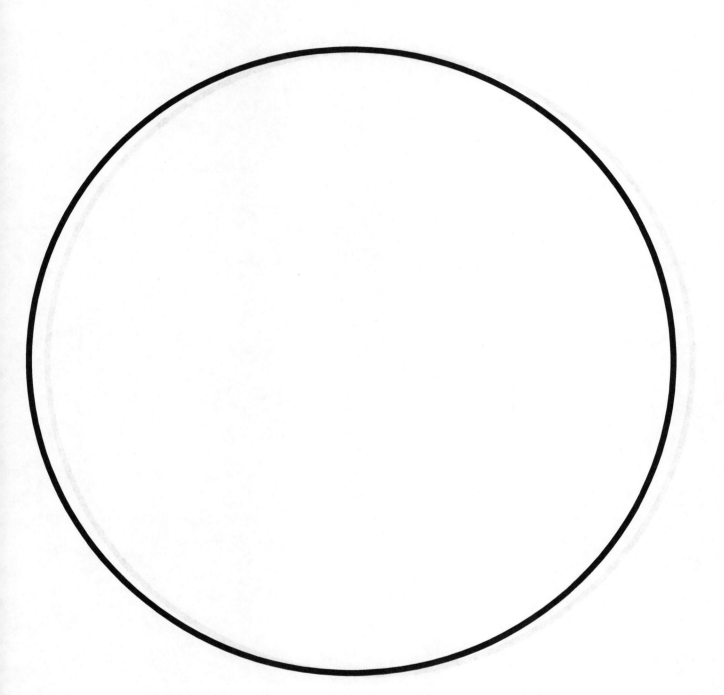

What is under my microscope?

_____

Date _____

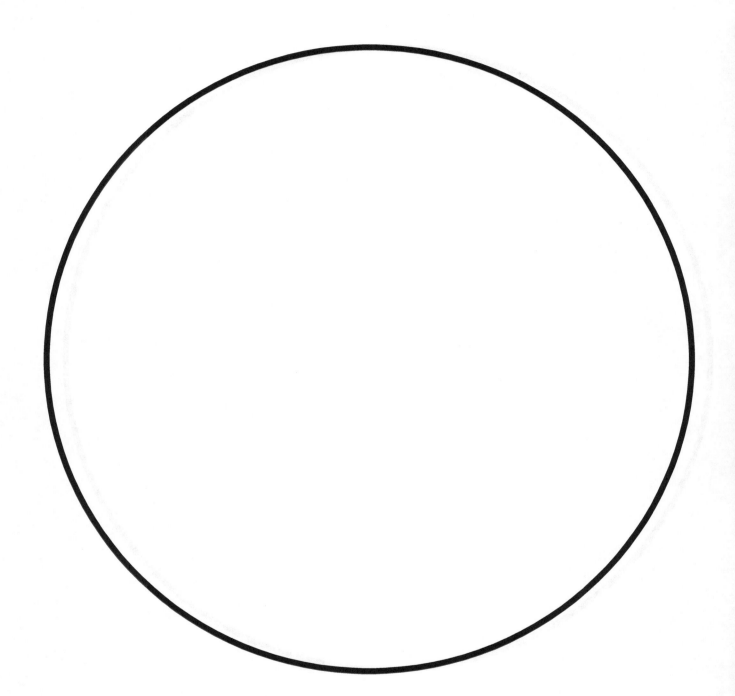

What is under my microscope?

_____

Date _____

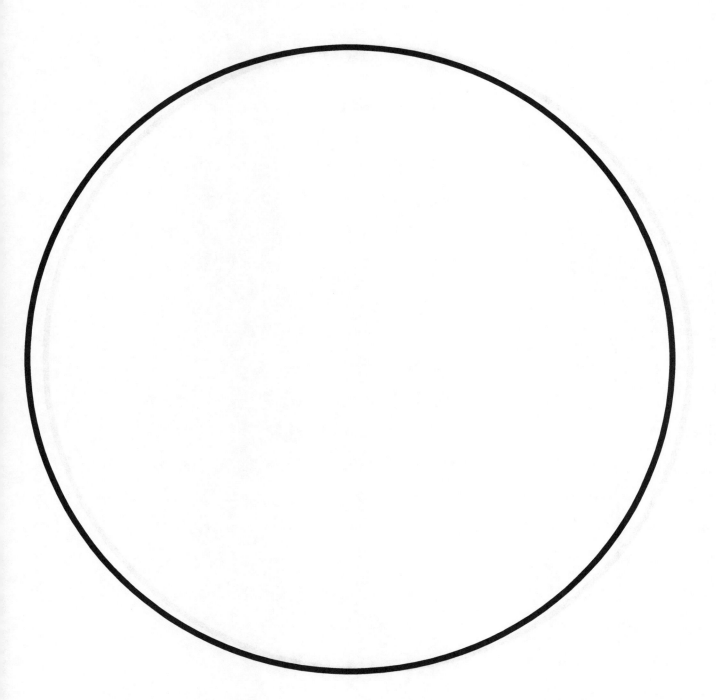

What is under my microscope?

_____

Date _____

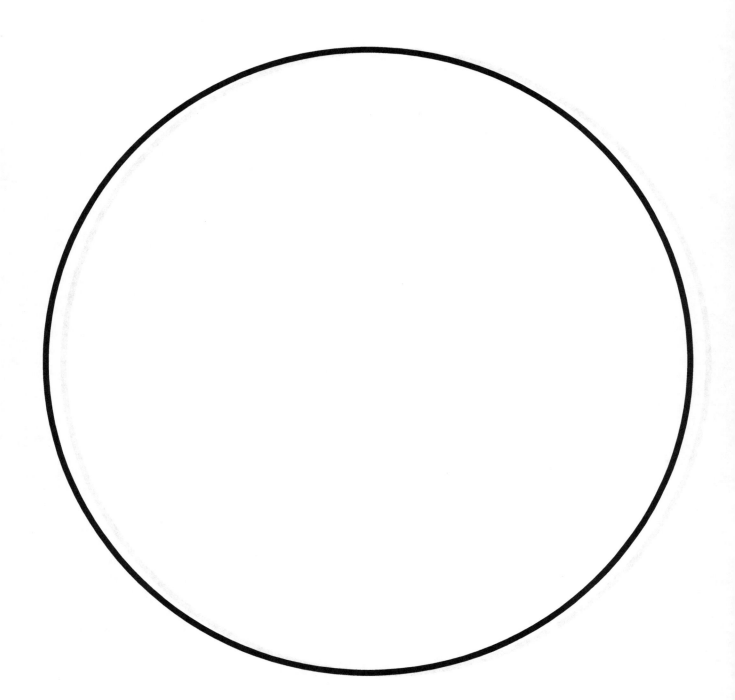

What is under my microscope?

_____

Date _____

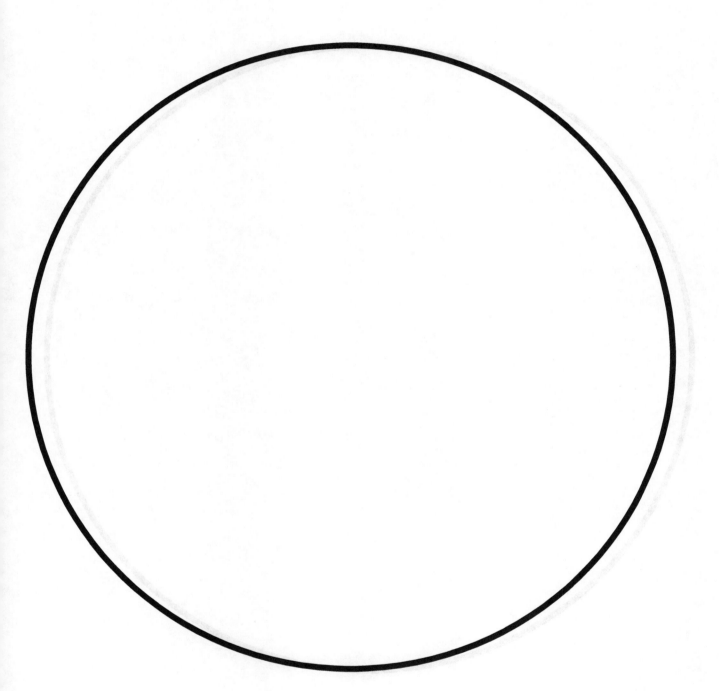

What is under my microscope?

_____

Date _____

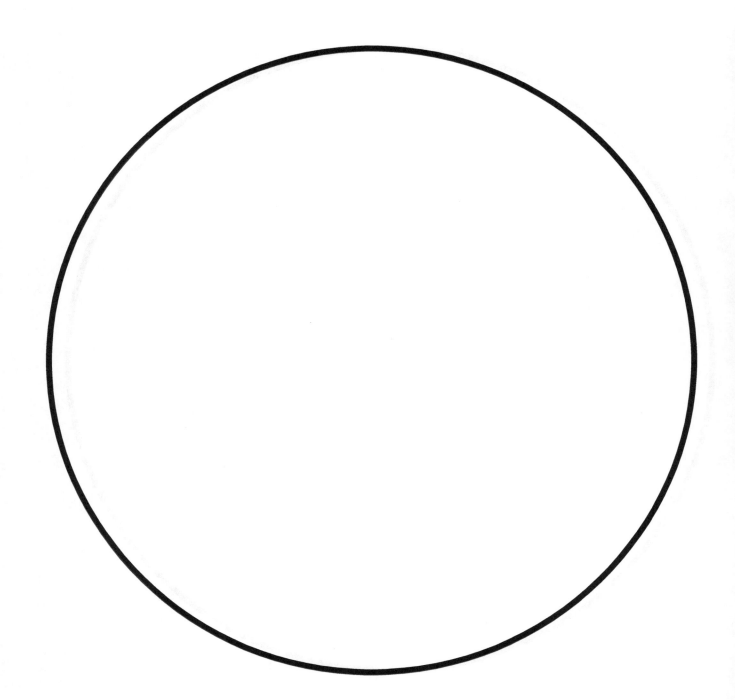

What is under my microscope?

_____

Date _____

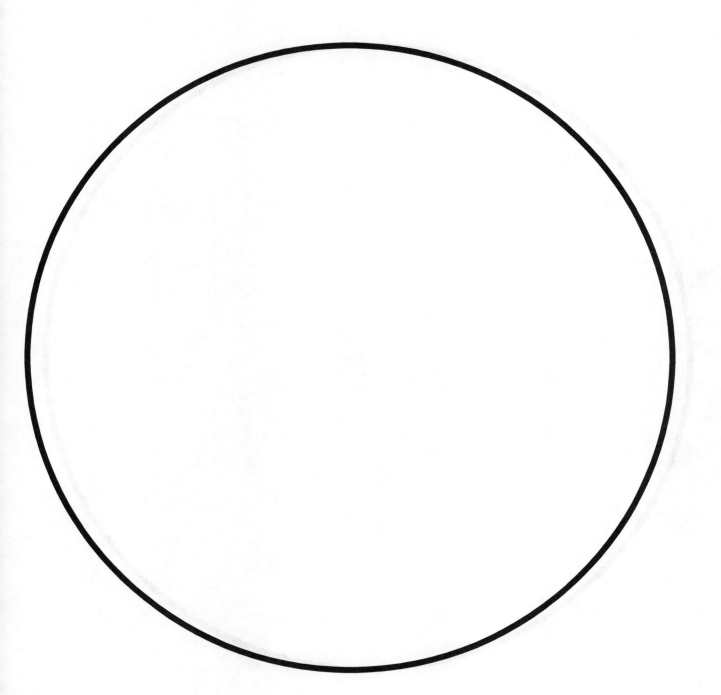

What is under my microscope?

_____

Date _____

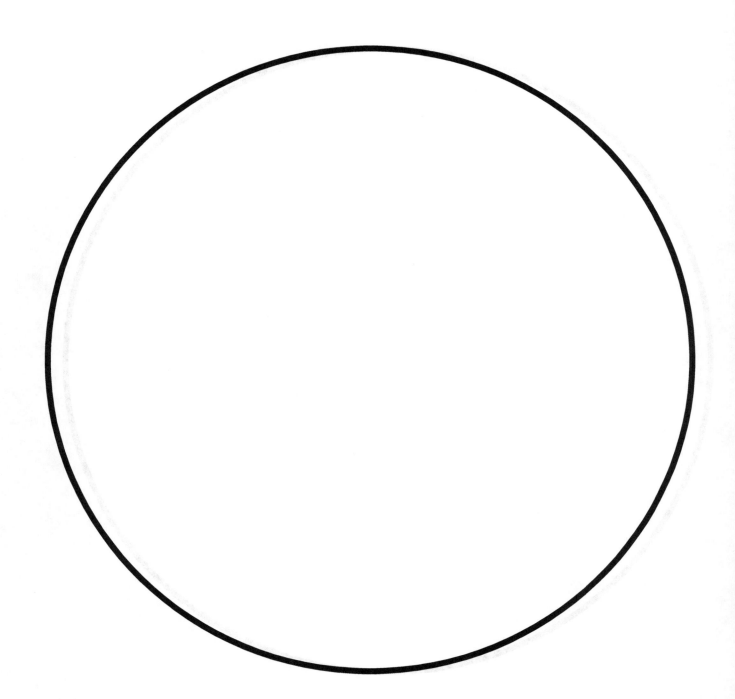

What is under my microscope?

_____

Date _____

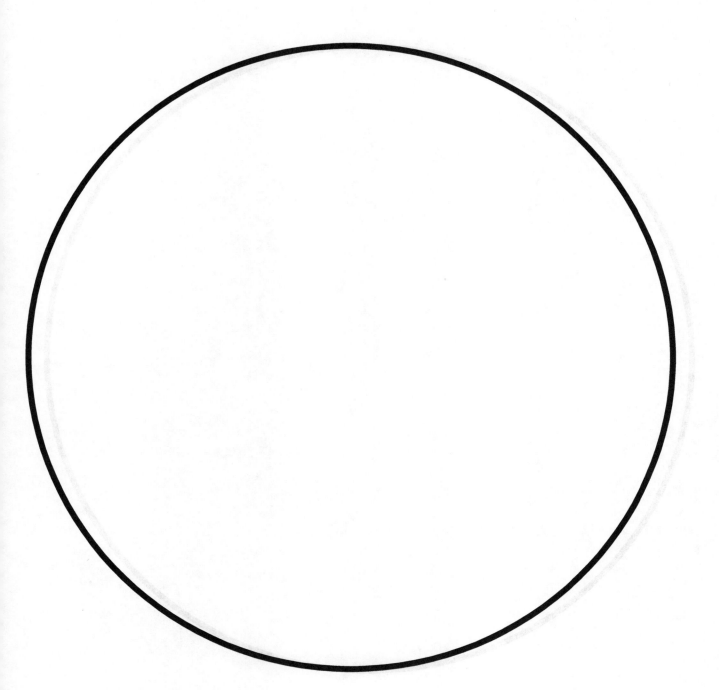

What is under my microscope?

_____

Date _____

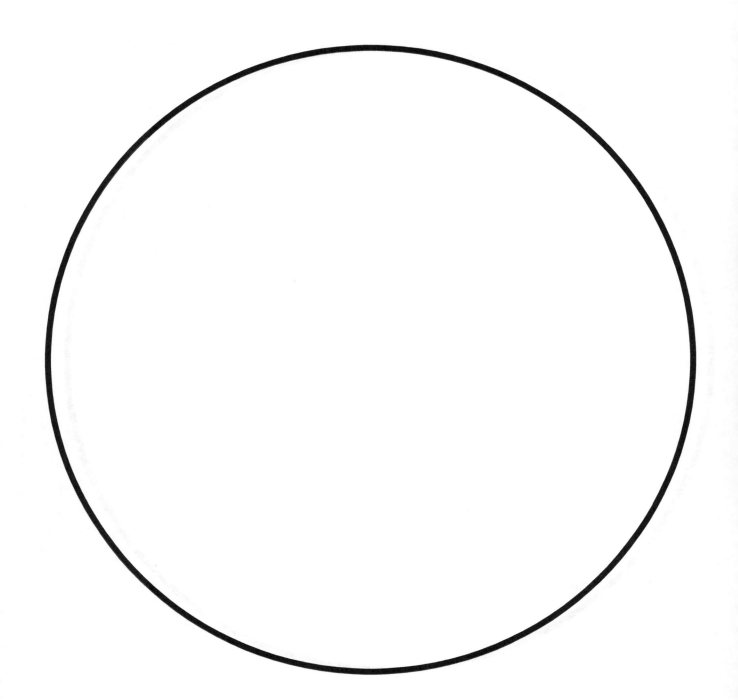

What is under my microscope?

_____

Date _____

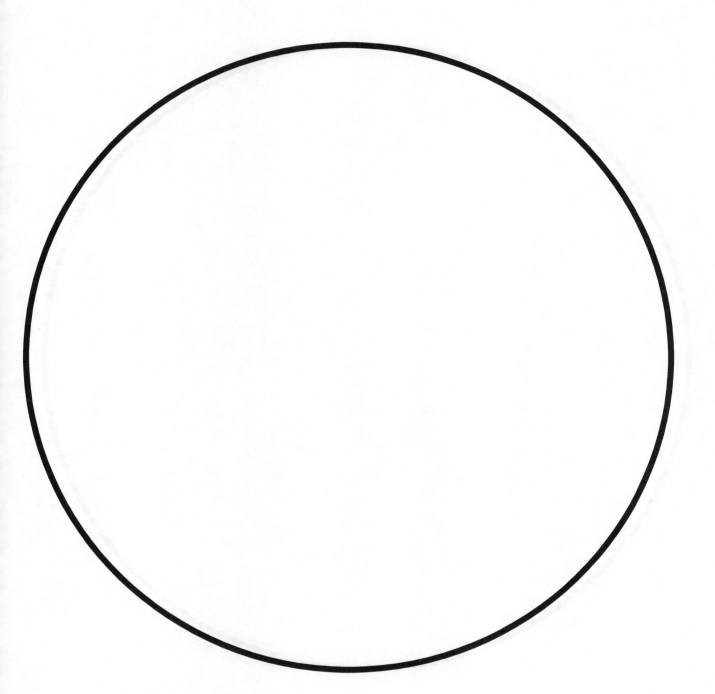

What is under my microscope?

_____

Date _____

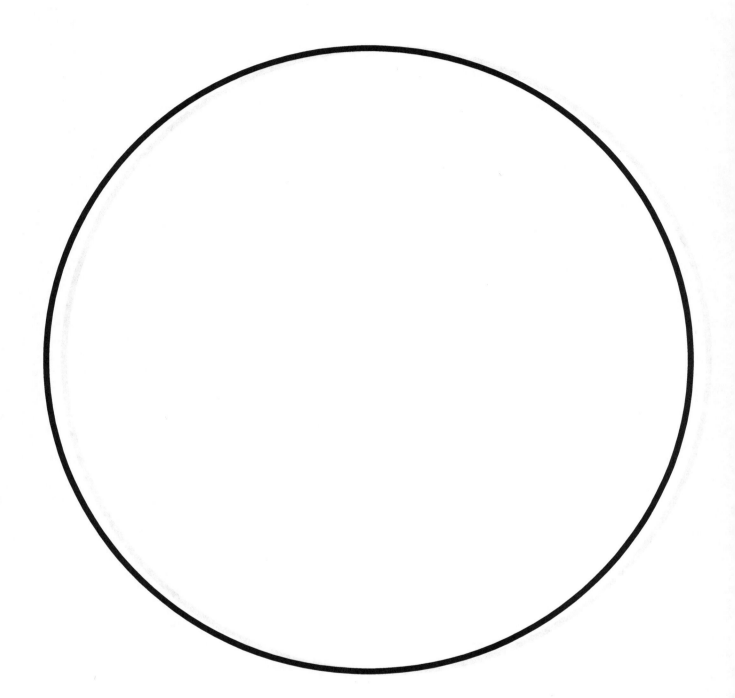

What is under my microscope?

_____

Date _____

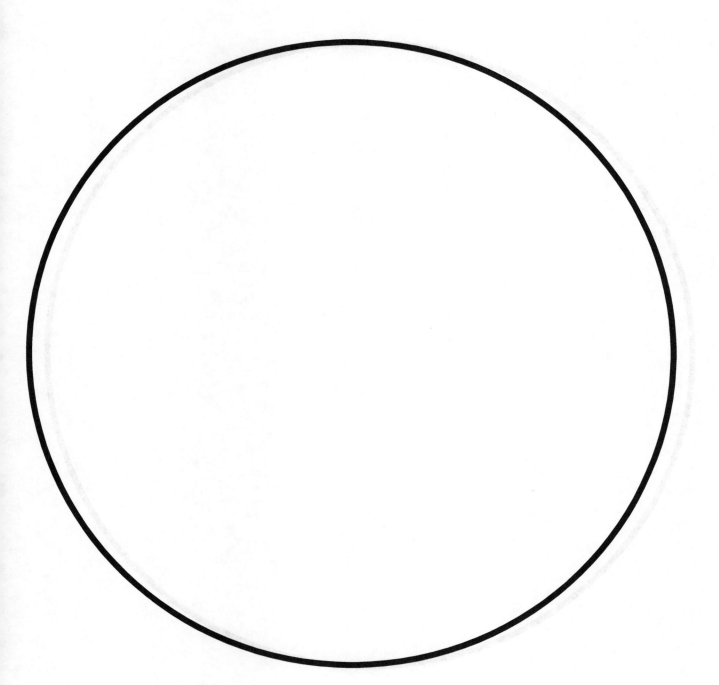

What is under my microscope?

_____

Date _____

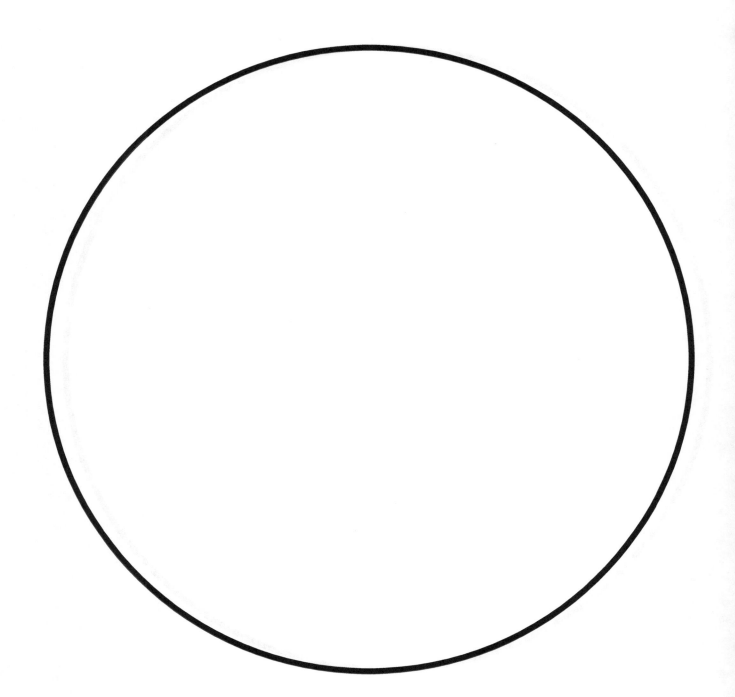

What is under my microscope?

_____

Date _____

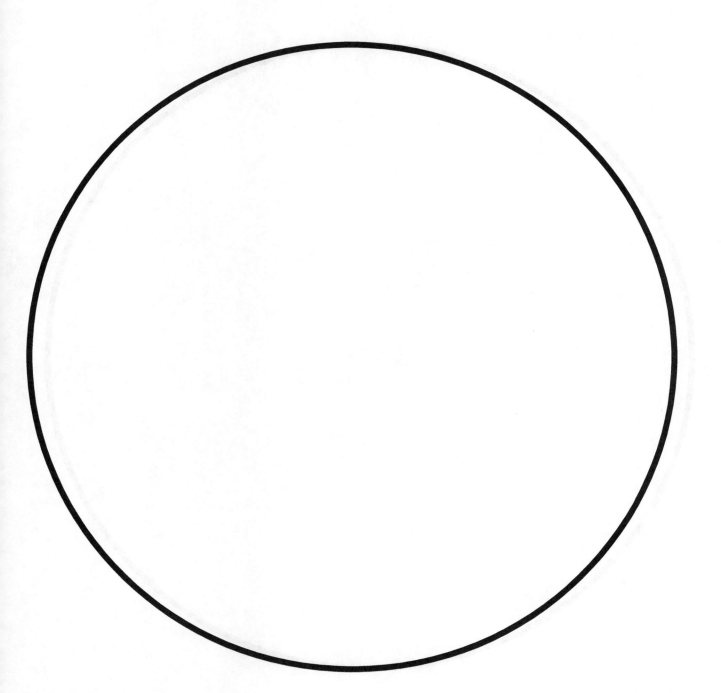

What is under my microscope?

_____

Date _____

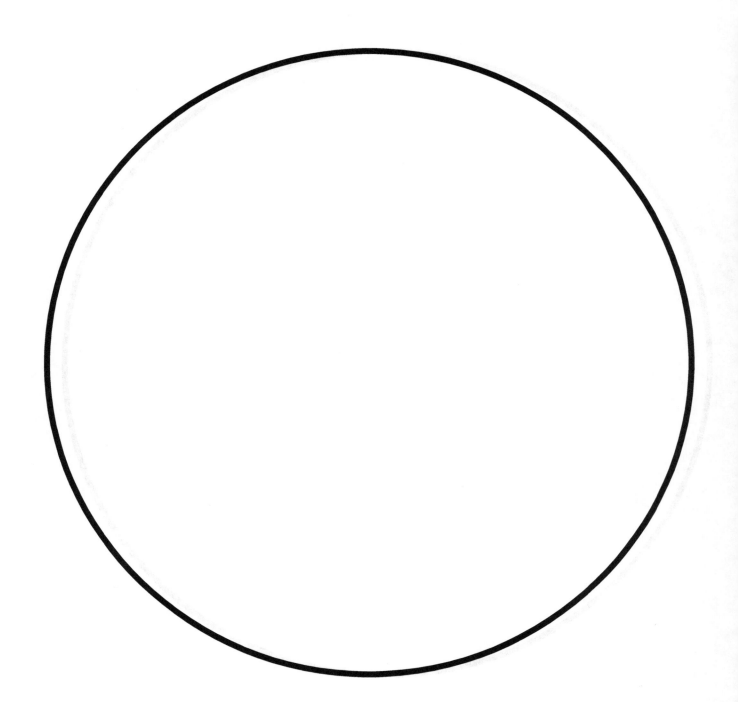

What is under my microscope?

_____

Date _____

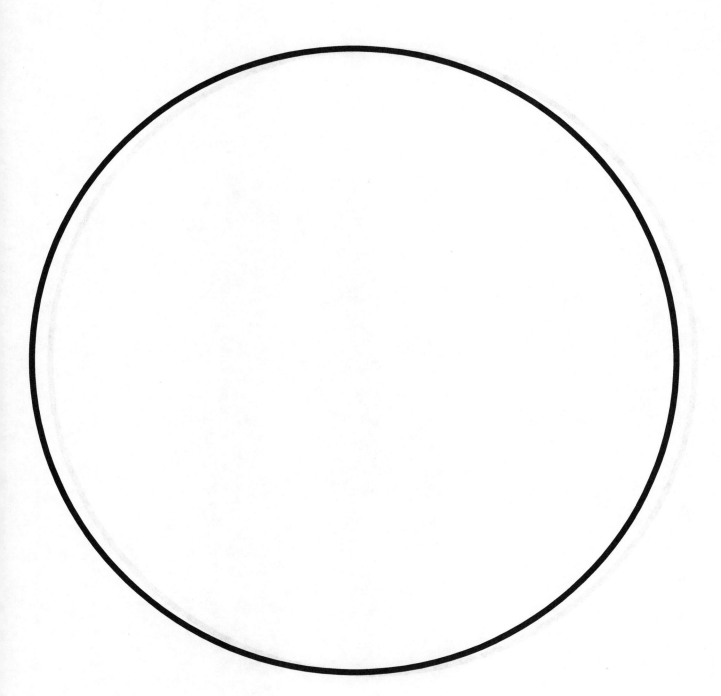

What is under my microscope?

_____

Date _____

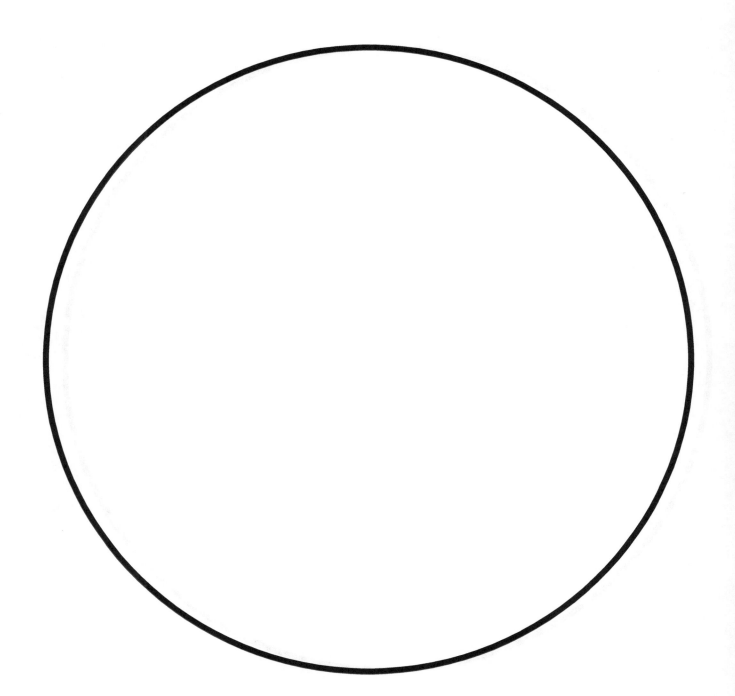

What is under my microscope?

_____

Date _____

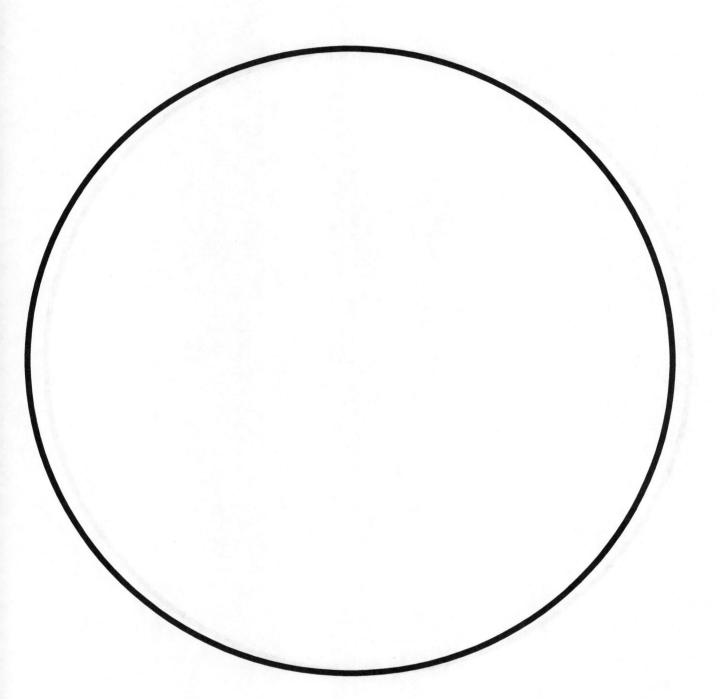

What is under my microscope?

_____

Date _____

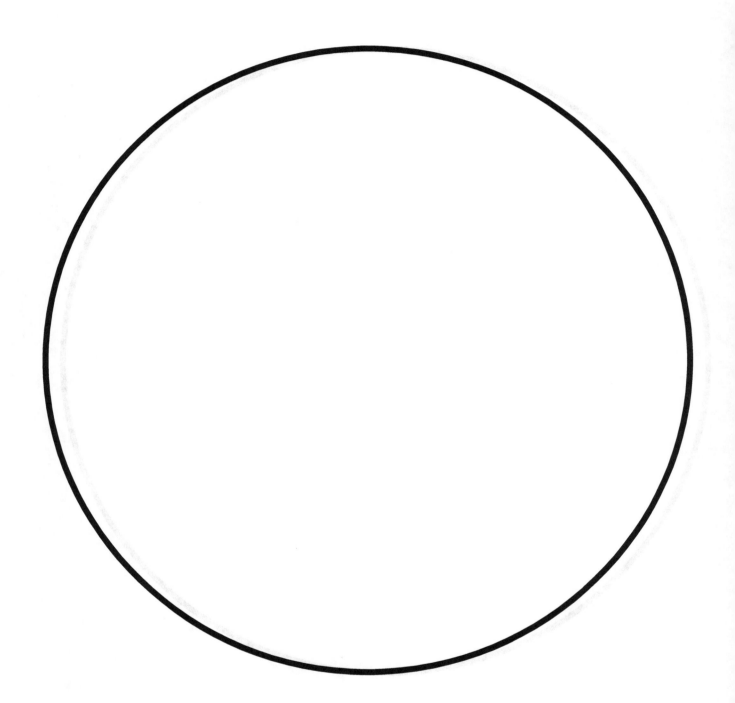

What is under my microscope?

_____

Date _____

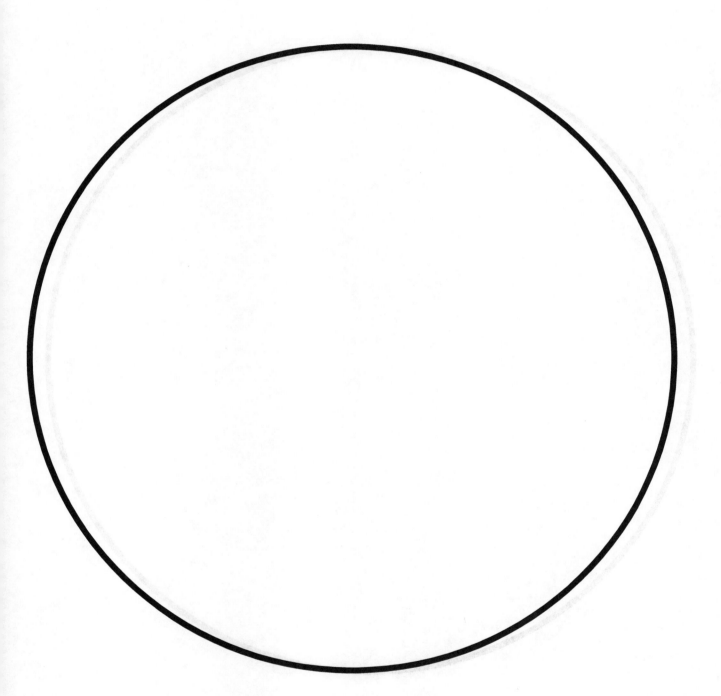

What is under my microscope?

_____

Date _____

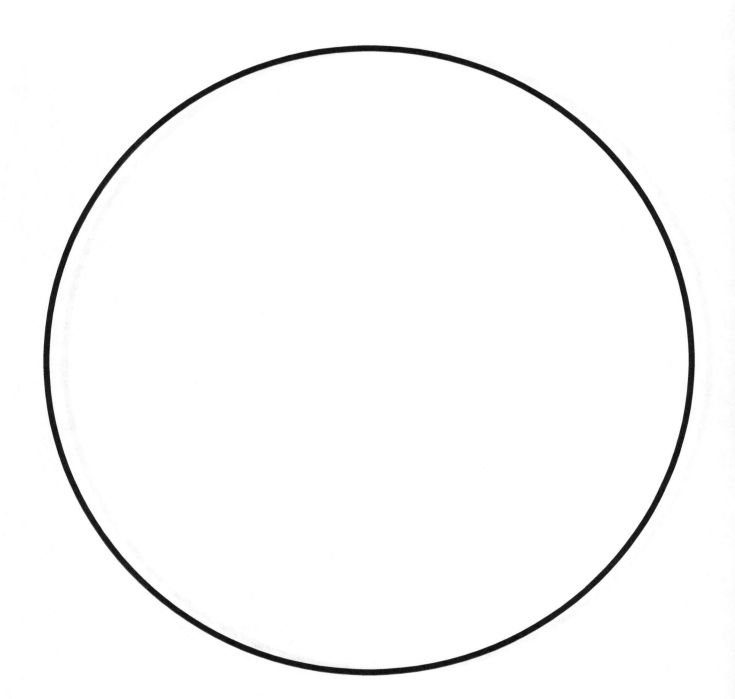

What is under my microscope?

_____

Date _____

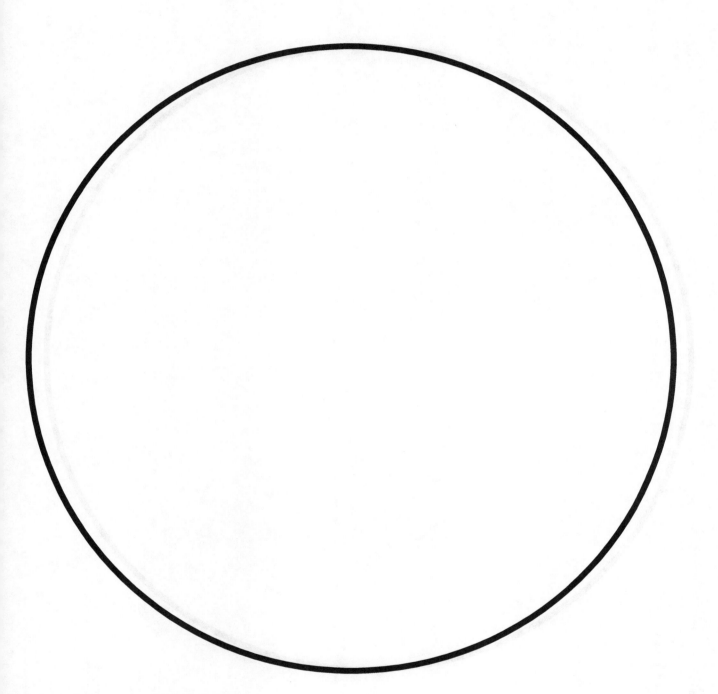

What is under my microscope?

_____

Date _____

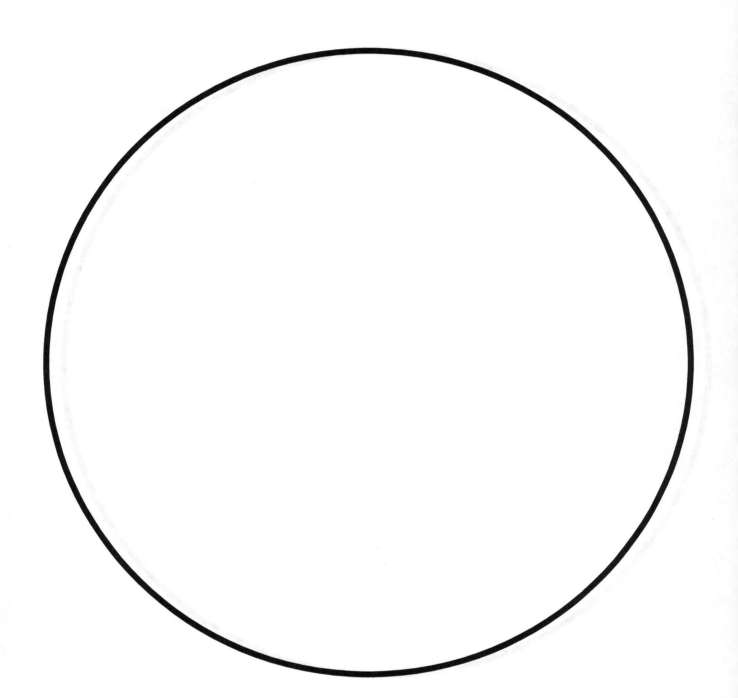

What is under my microscope?

_____

Date _____

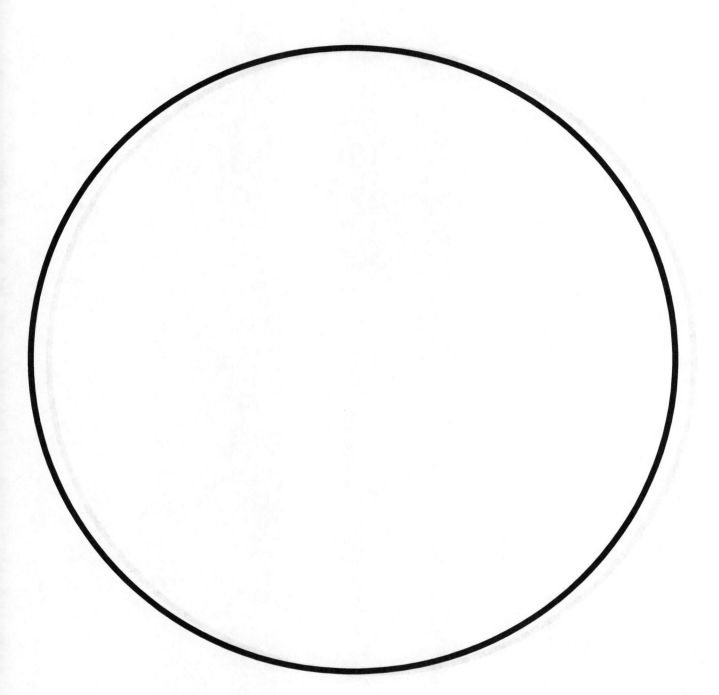

What is under my microscope?

_____

Date _____

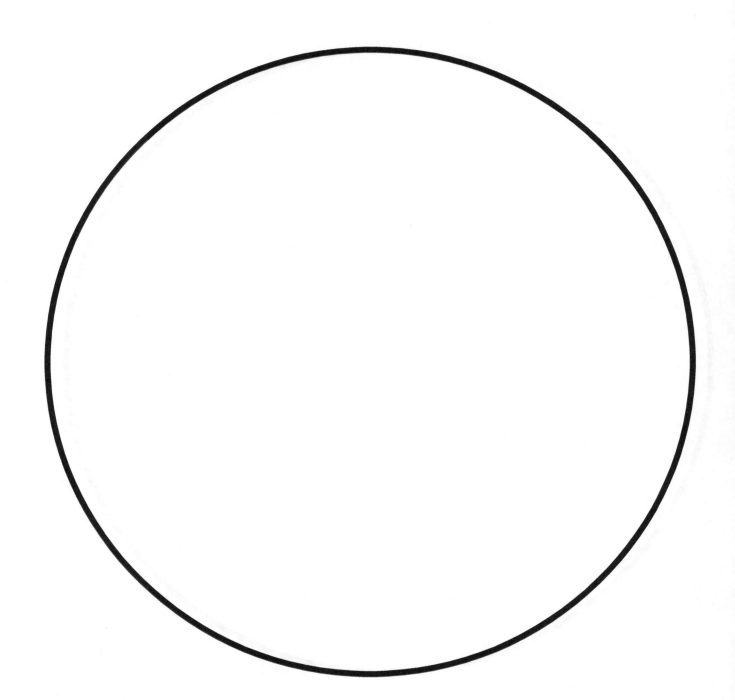

What is under my microscope?

_____

Date _____

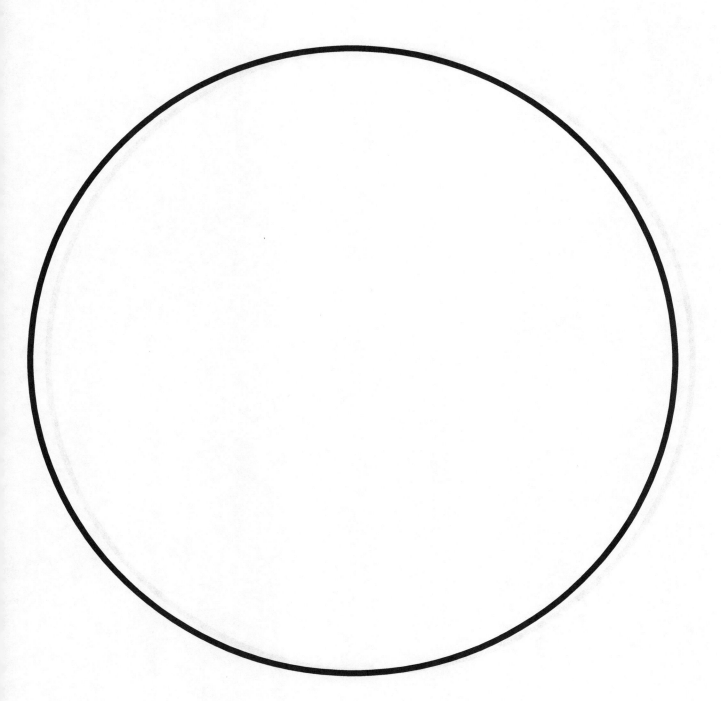

What is under my microscope?

_____

Date _____

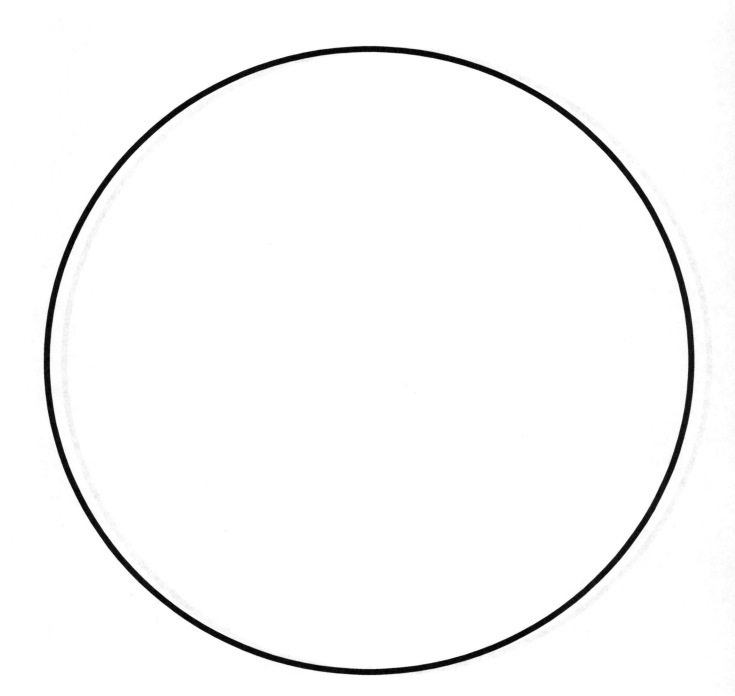

What is under my microscope?

_____

Date _____

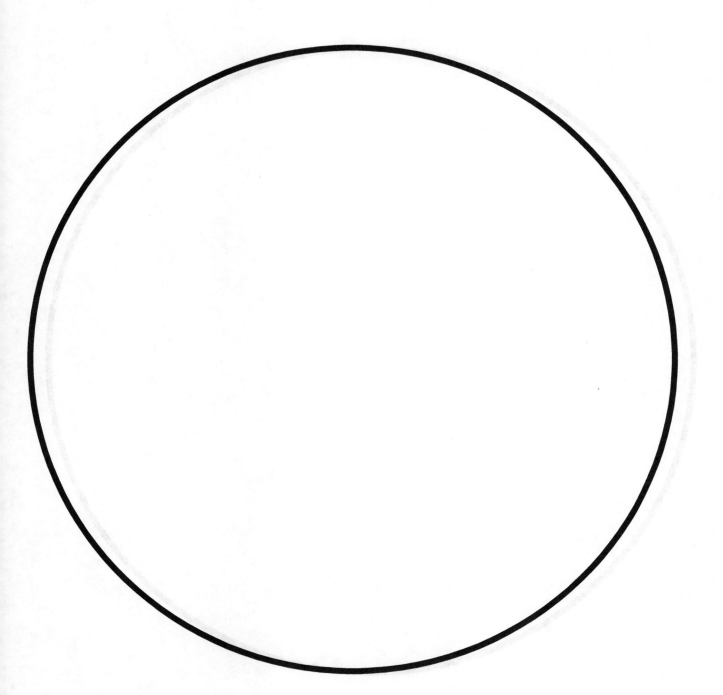

What is under my microscope?

_____

Date _____

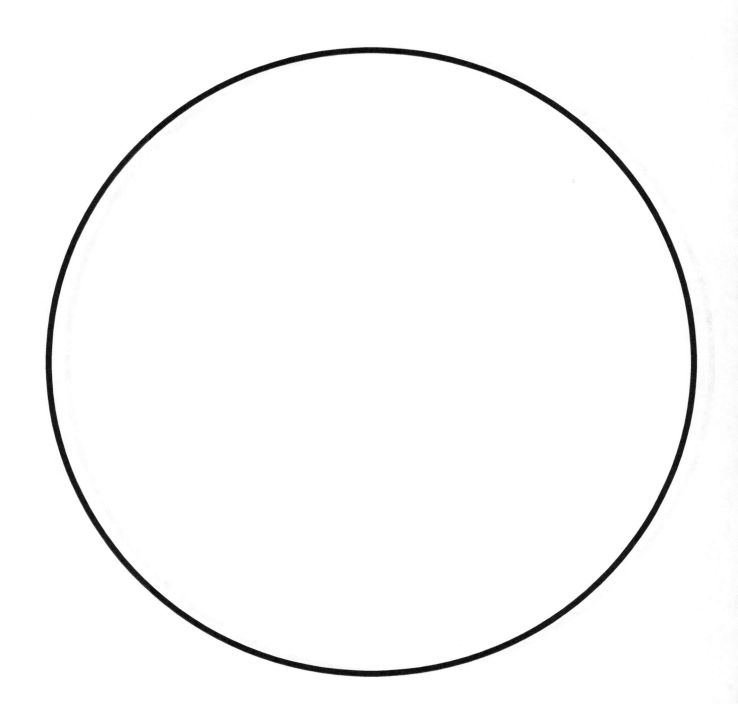

What is under my microscope?

_____

Date _____

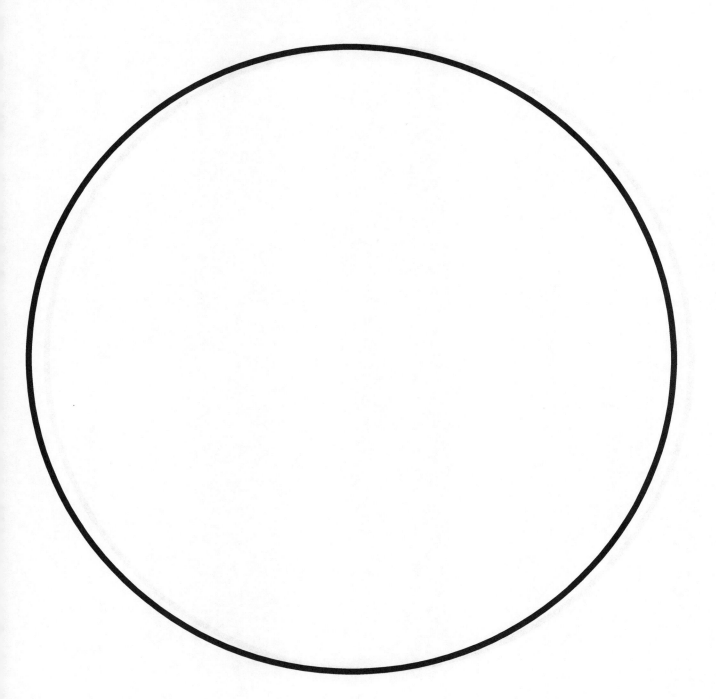

What is under my microscope?

_____

Date _____

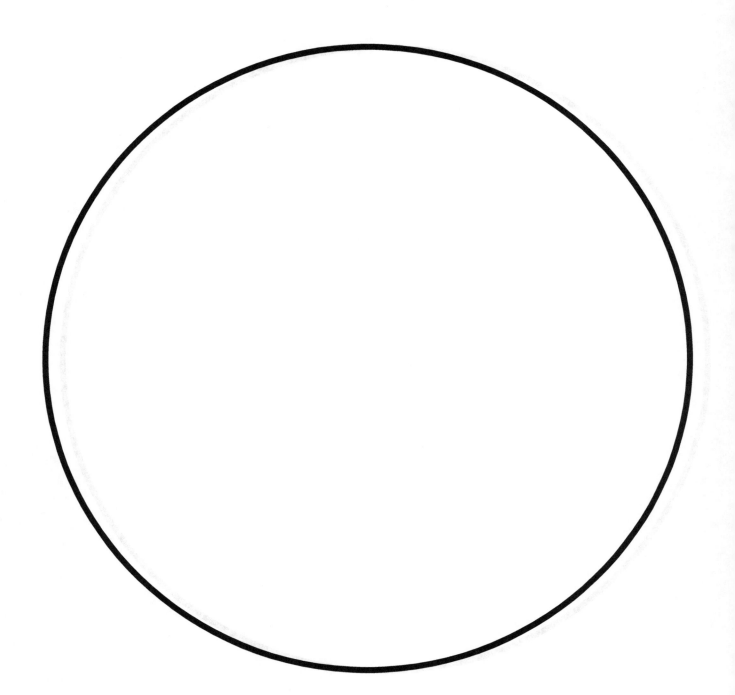

What is under my microscope?

_____

Date _____

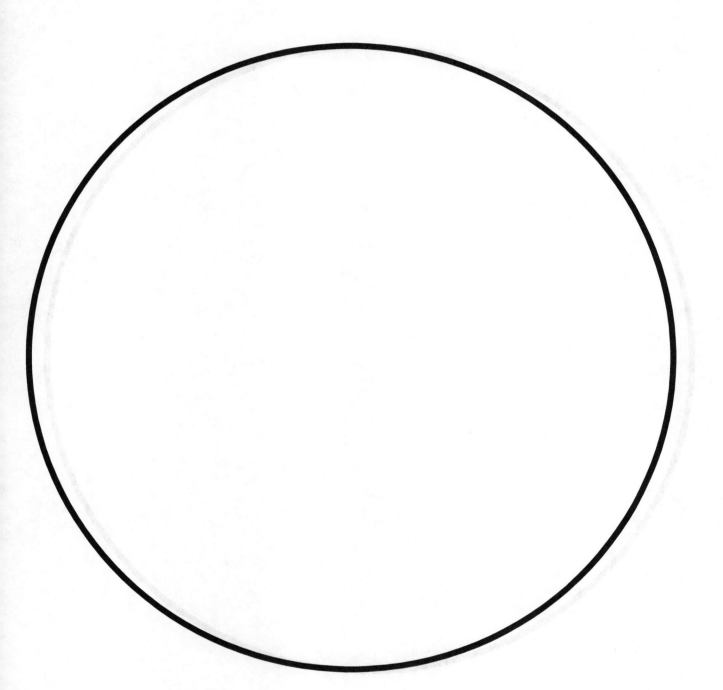

What is under my microscope?

_____

Date _____

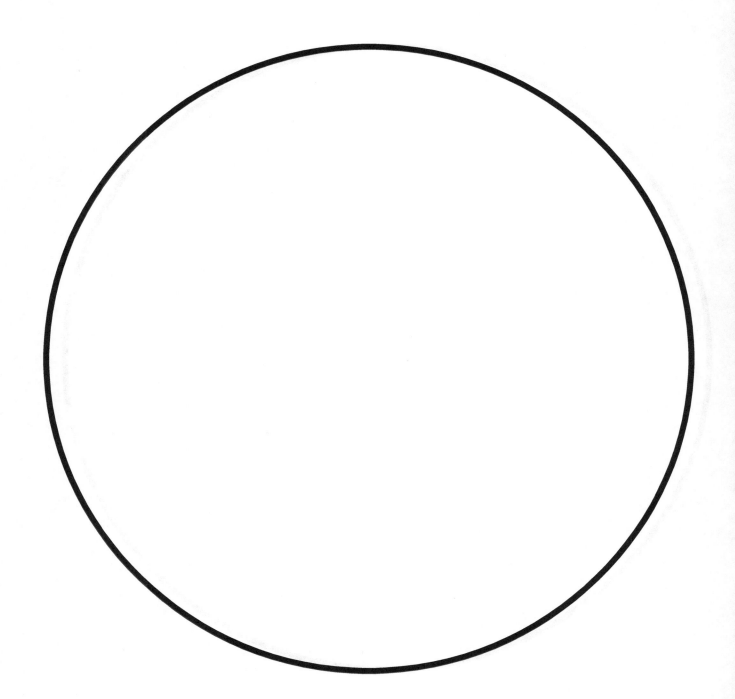

What is under my microscope?

_____

Date _____

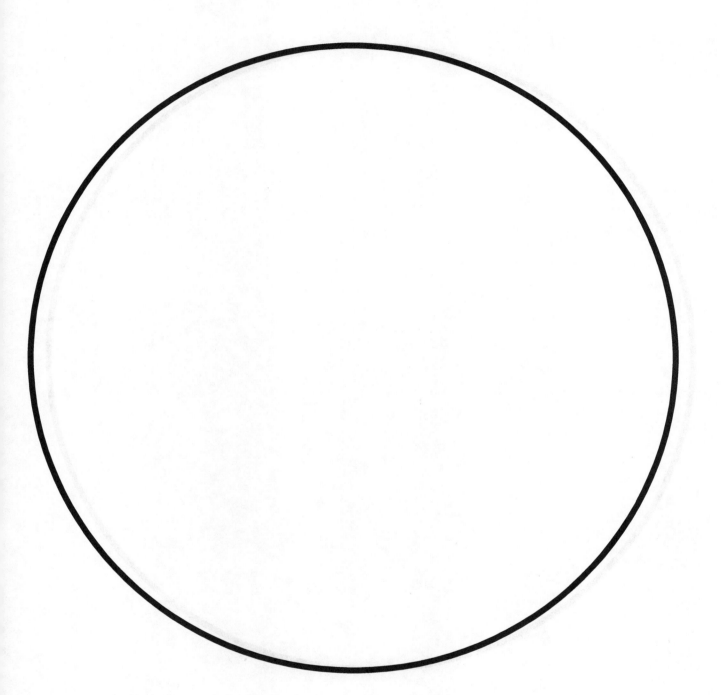

What is under my microscope?

_____

Date _____

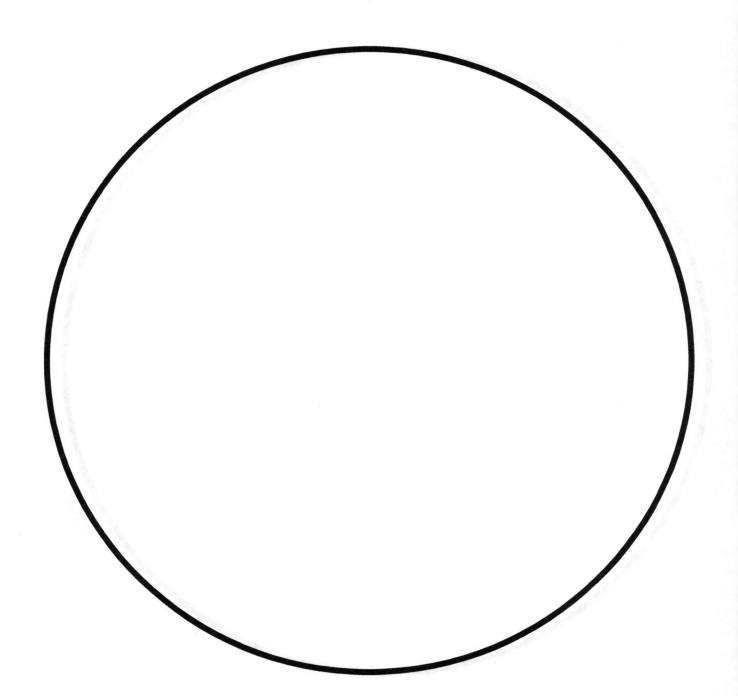

What is under my microscope?

_____

Date _____

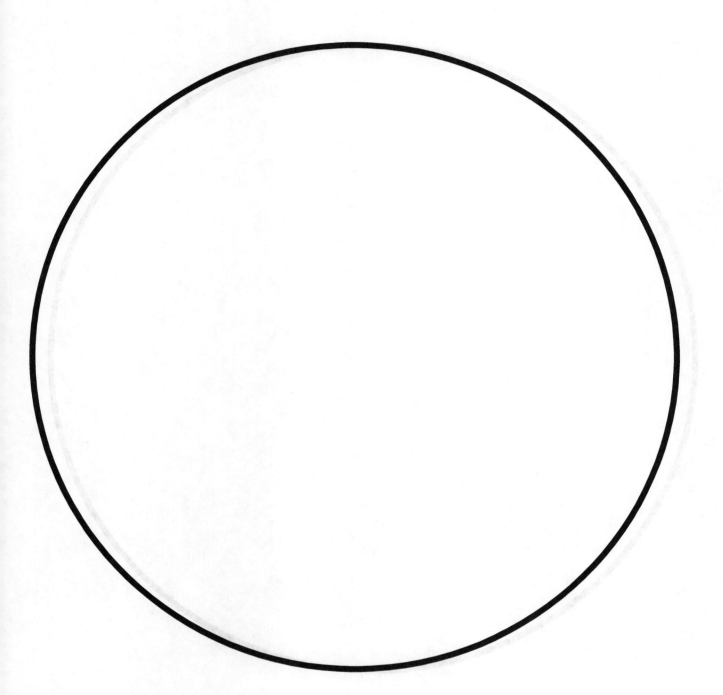

What is under my microscope?

_____

Date _____

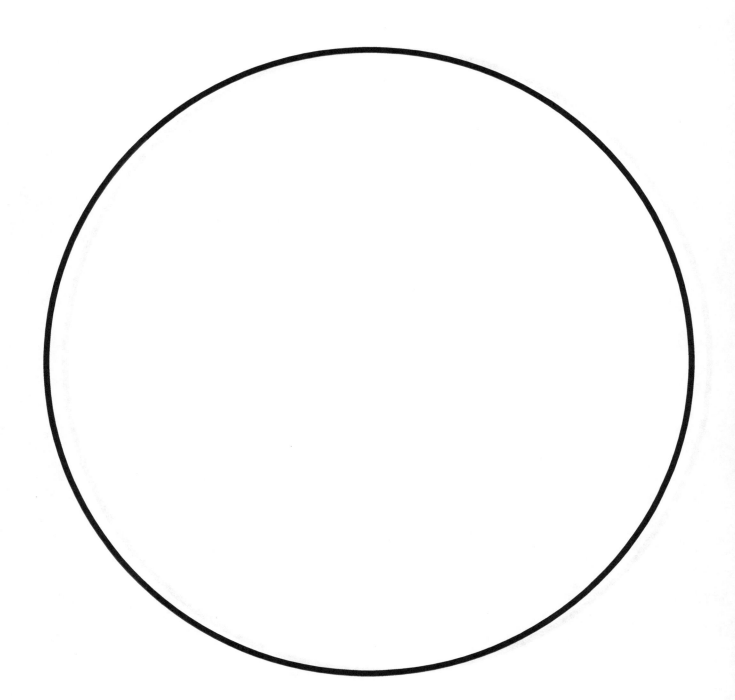

What is under my microscope?

_____

Date _____

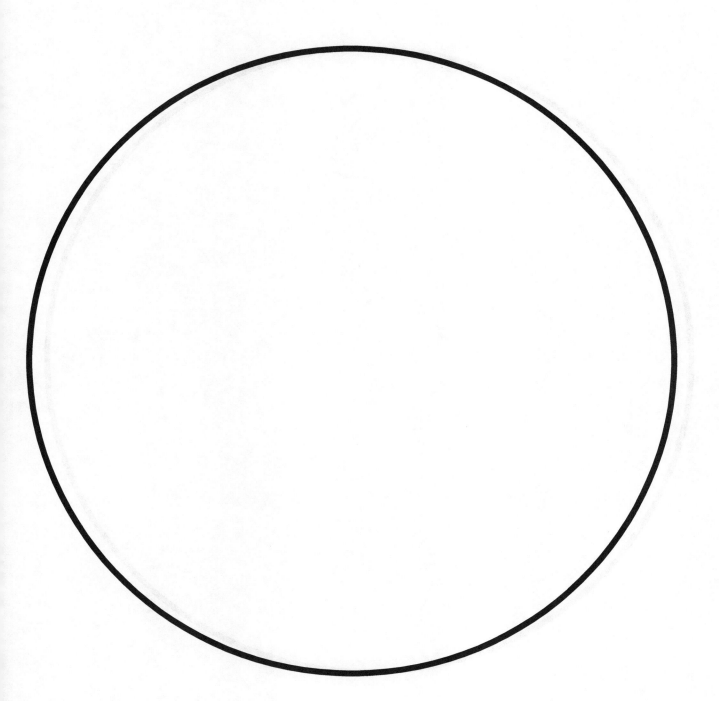

What is under my microscope?

_____

Date _____

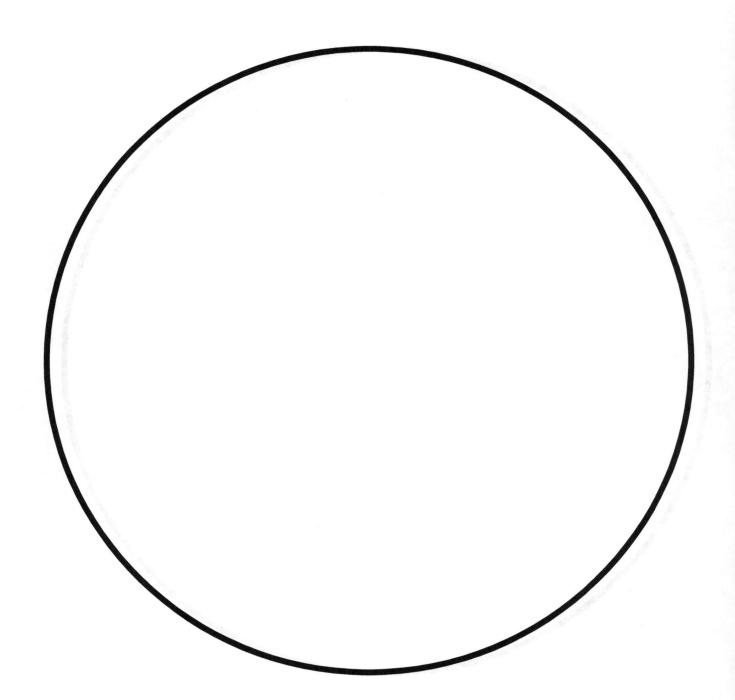

What is under my microscope?

_____

Date _____

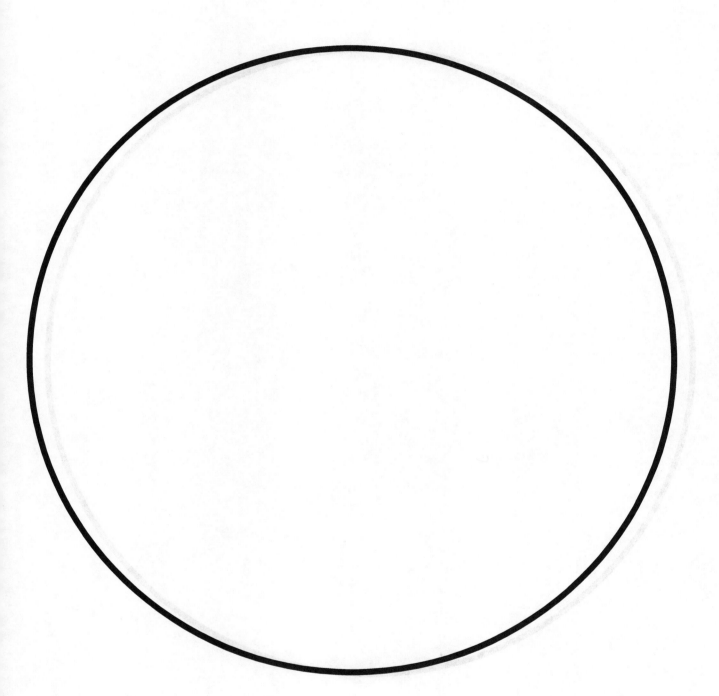

What is under my microscope?

_____

Date _____

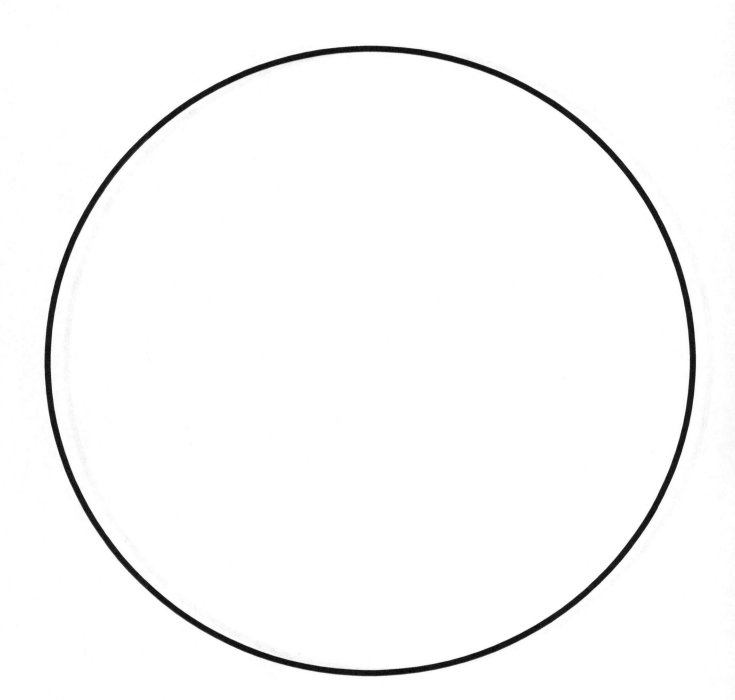

What is under my microscope?

_____

Date _____

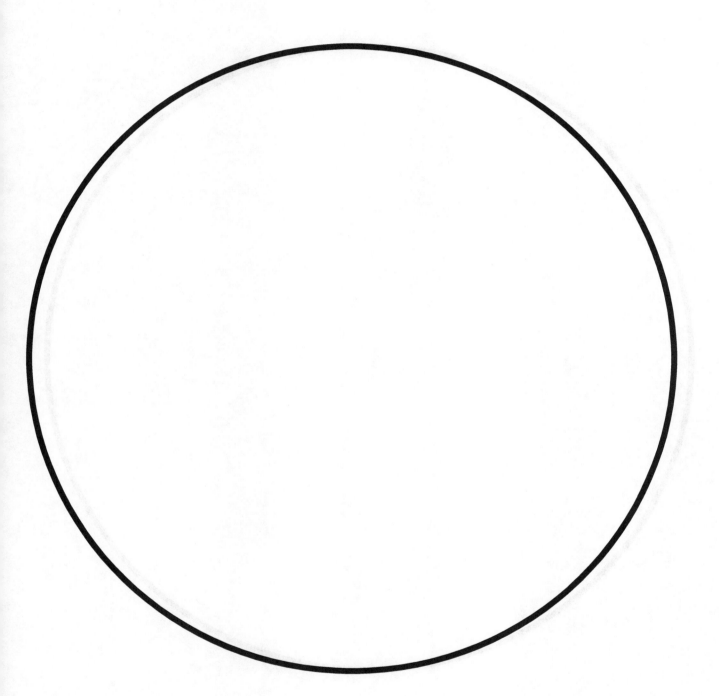

What is under my microscope?

_____

Date _____

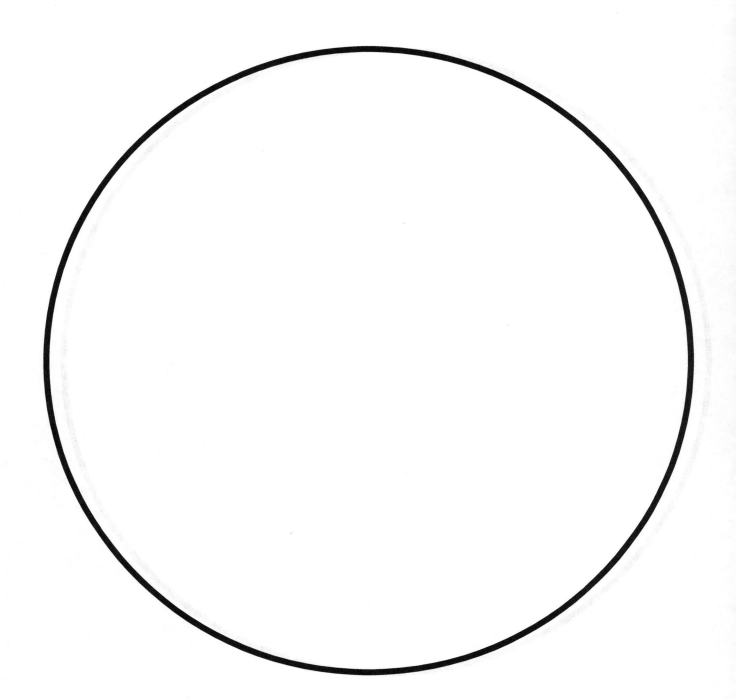

What is under my microscope?

_____

Date _____

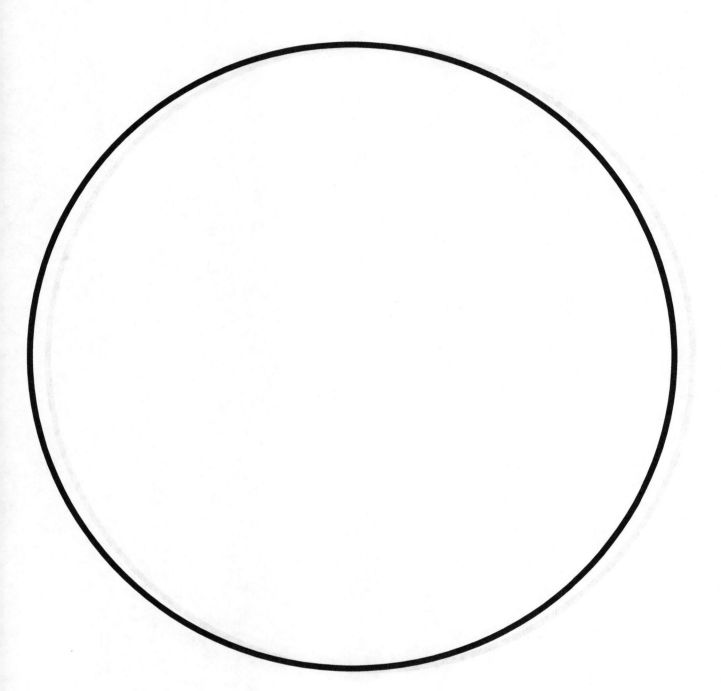

What is under my microscope?

_____

Date _____

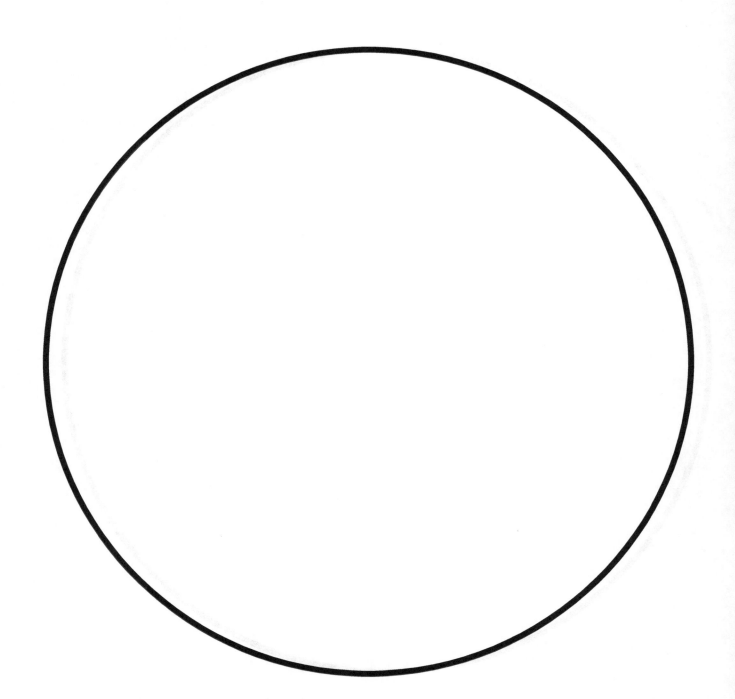

What is under my microscope?

_____

Date _____

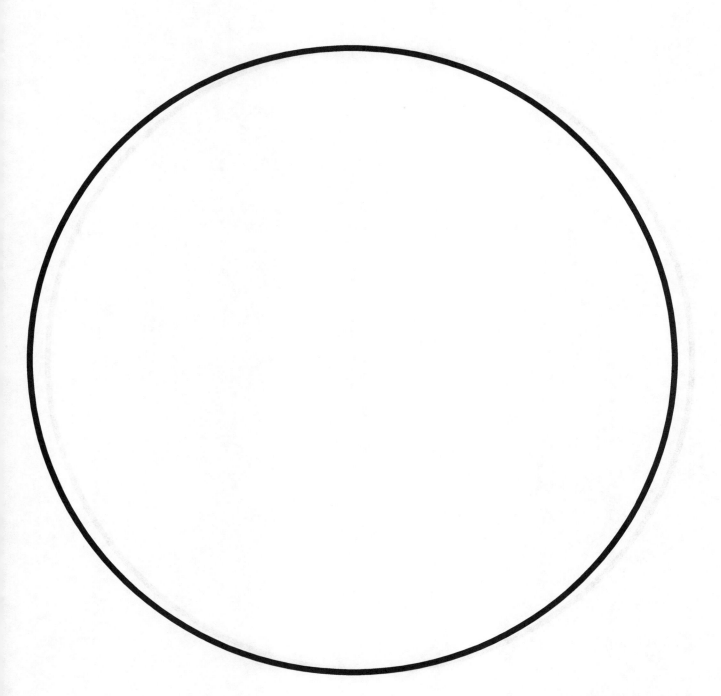

What is under my microscope?

_____

Date _____

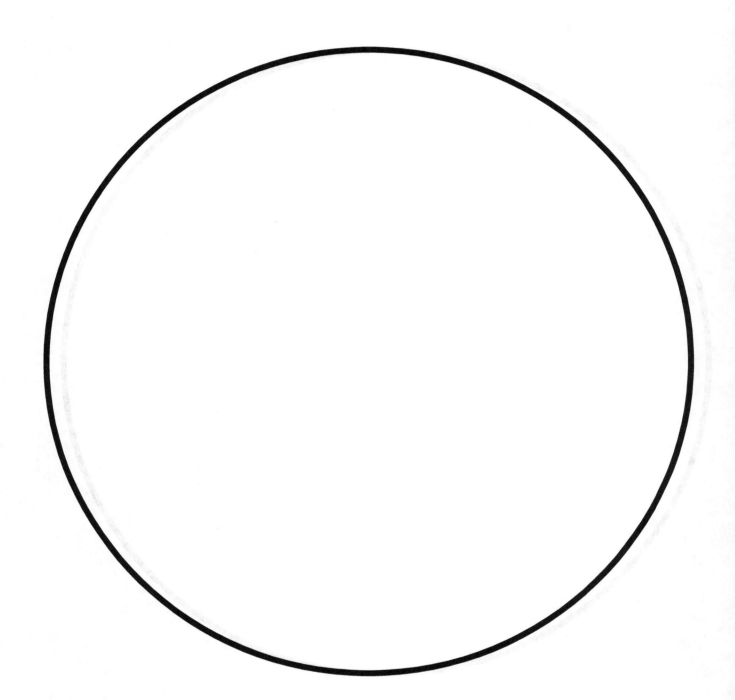

What is under my microscope?

_____

Date _____

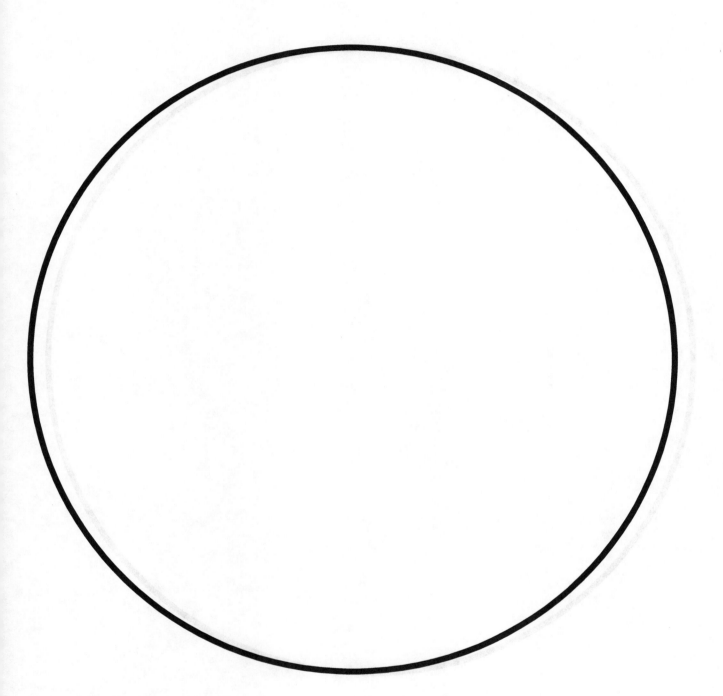

What is under my microscope?

_____

Date _____

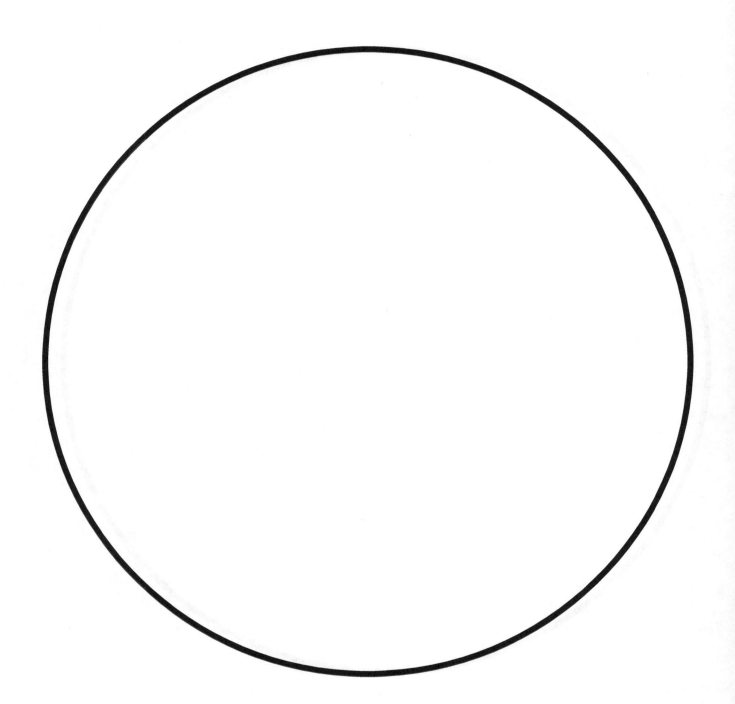

What is under my microscope?

_____

Date _____

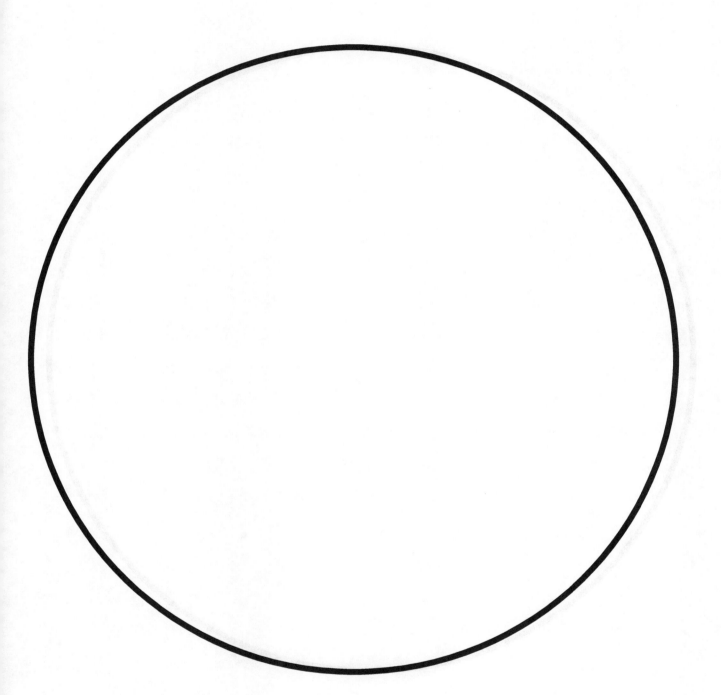

What is under my microscope?

_____

Date _____

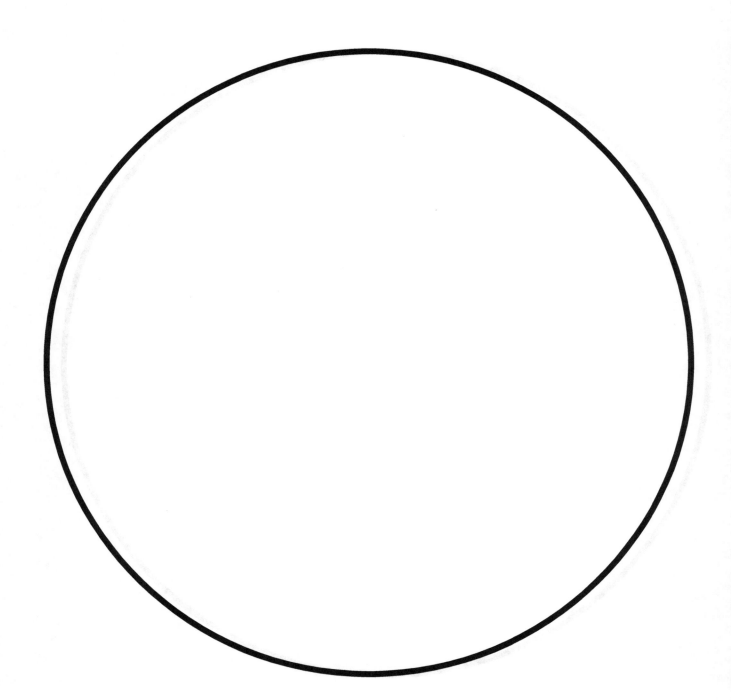

What is under my microscope?

_____

Date _____

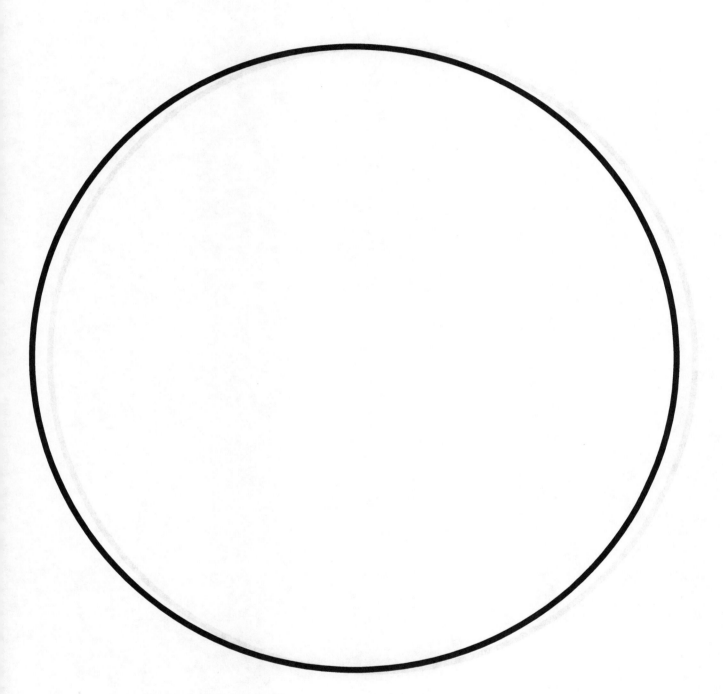

What is under my microscope?

_____

Date _____

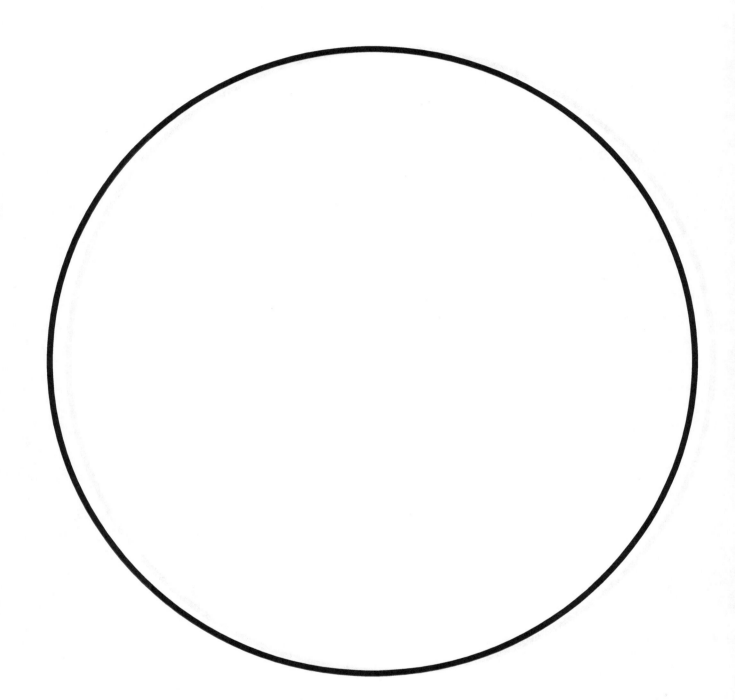

What is under my microscope?

_____

Date _____

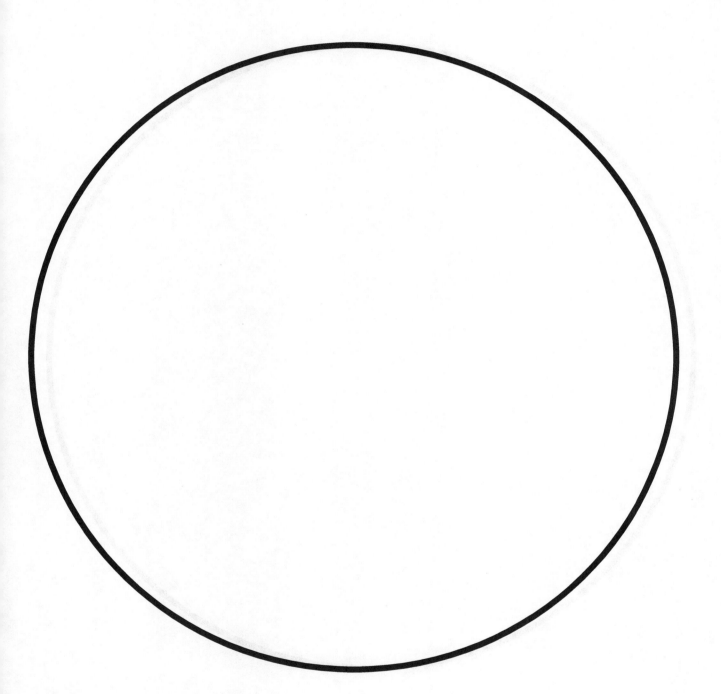

What is under my microscope?

_____

Date _____

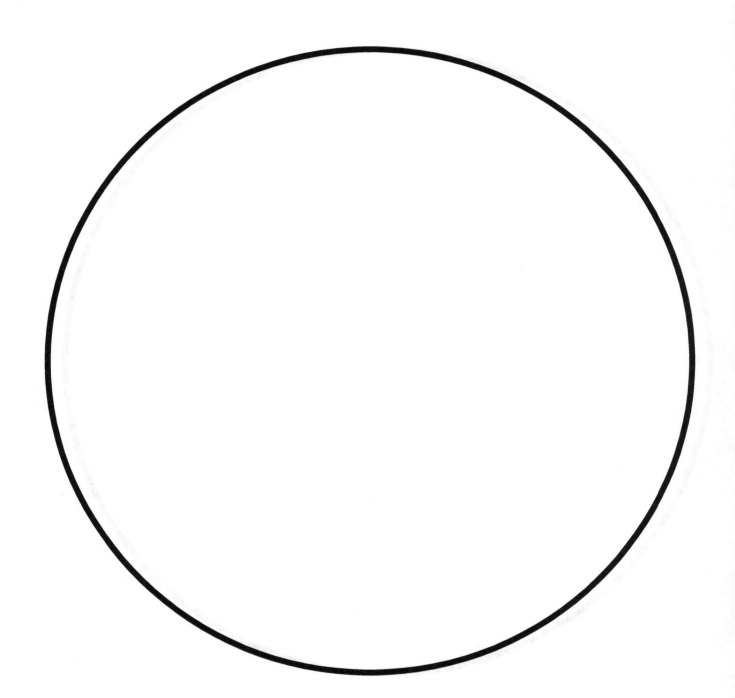

What is under my microscope?

_____

Date _____

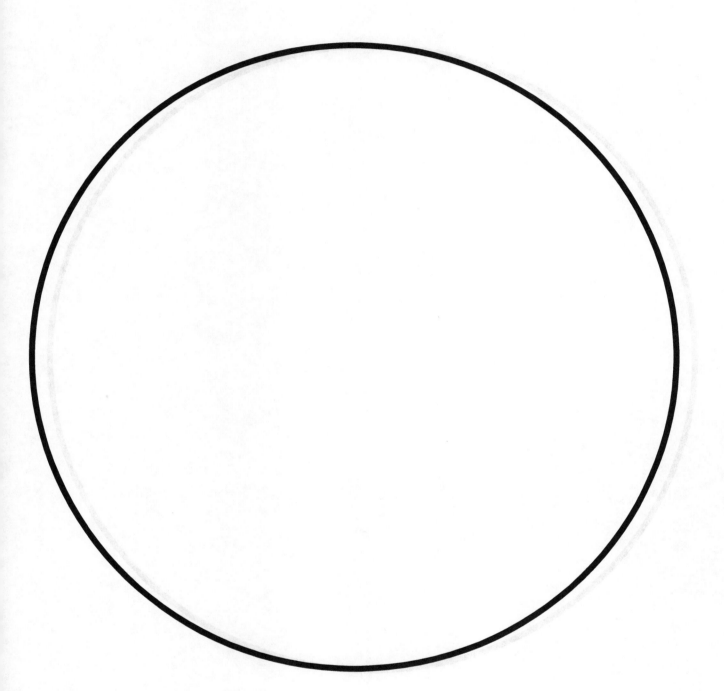

What is under my microscope?

_____

Date _____

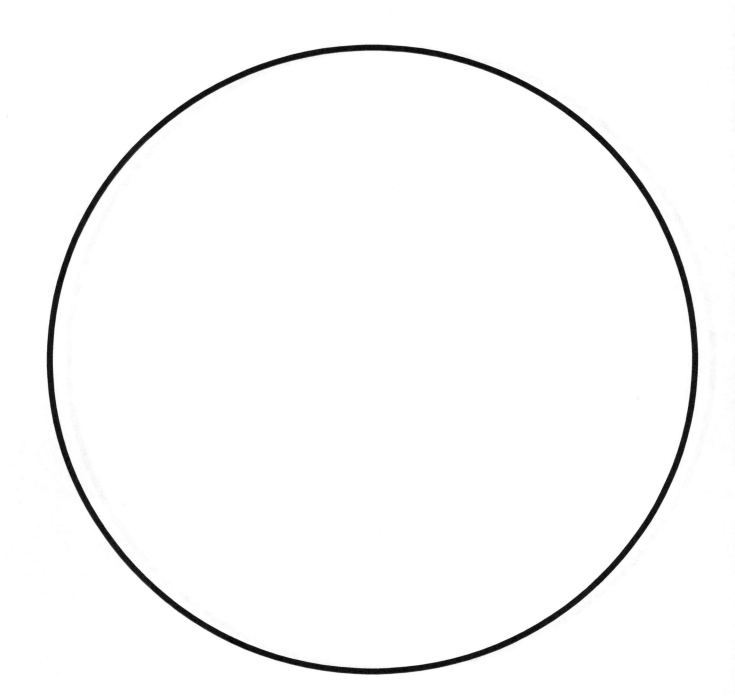

What is under my microscope?

_____

Date _____

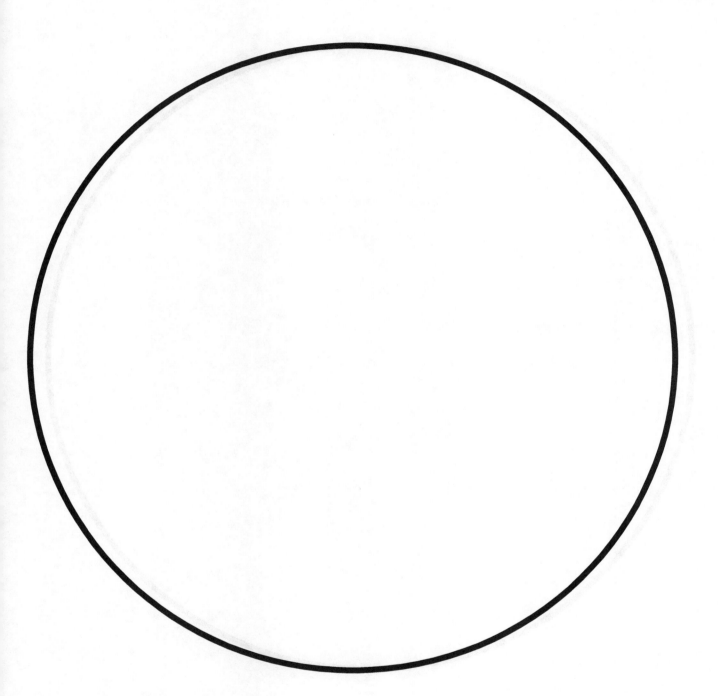

What is under my microscope?

_____

Date _____

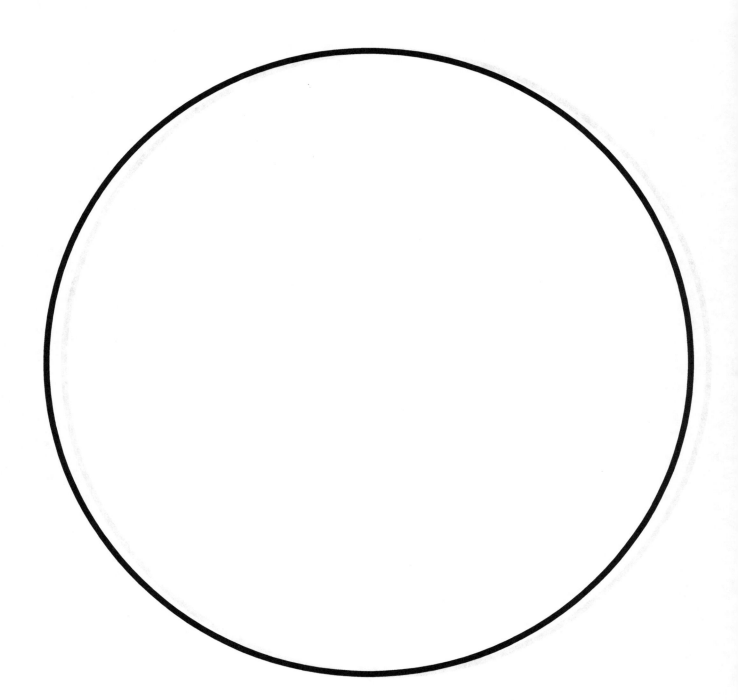

What is under my microscope?

_____

Date _____

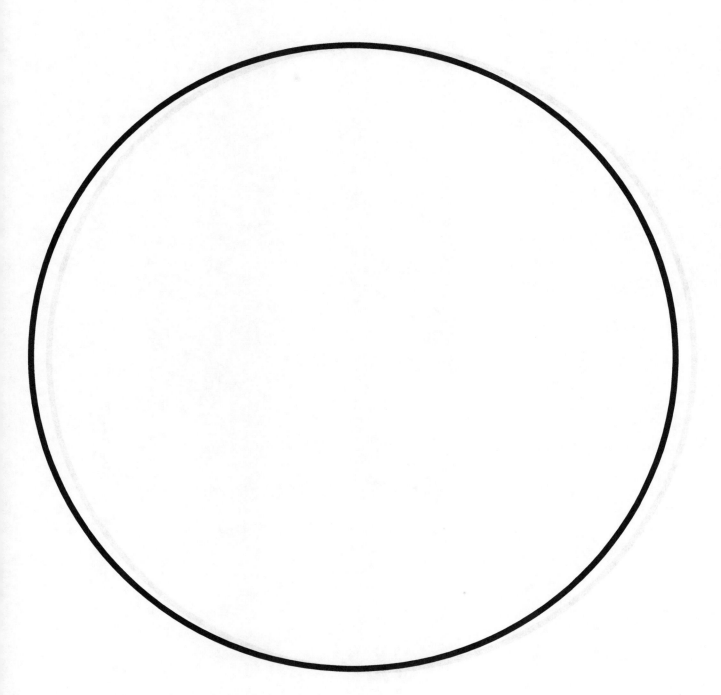

What is under my microscope?

_____

Date _____

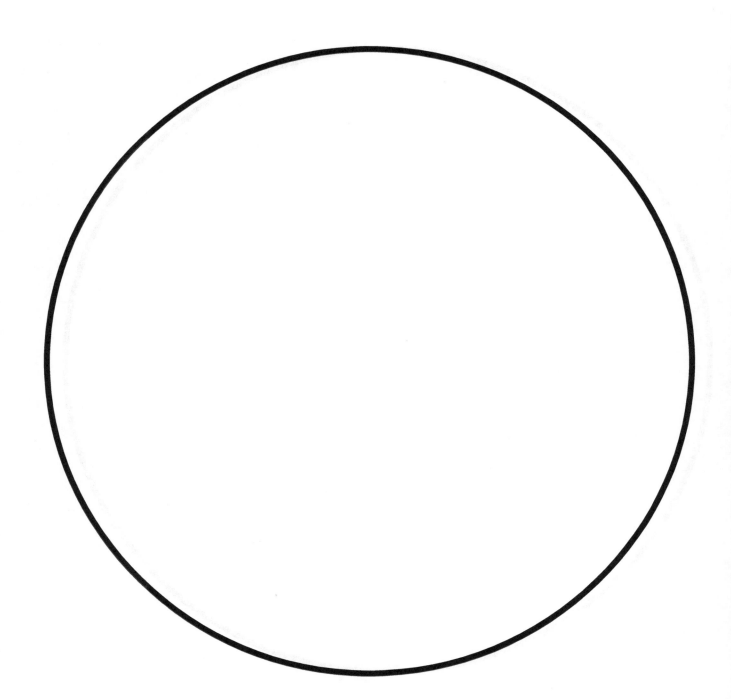

What is under my microscope?

_____

Date _____

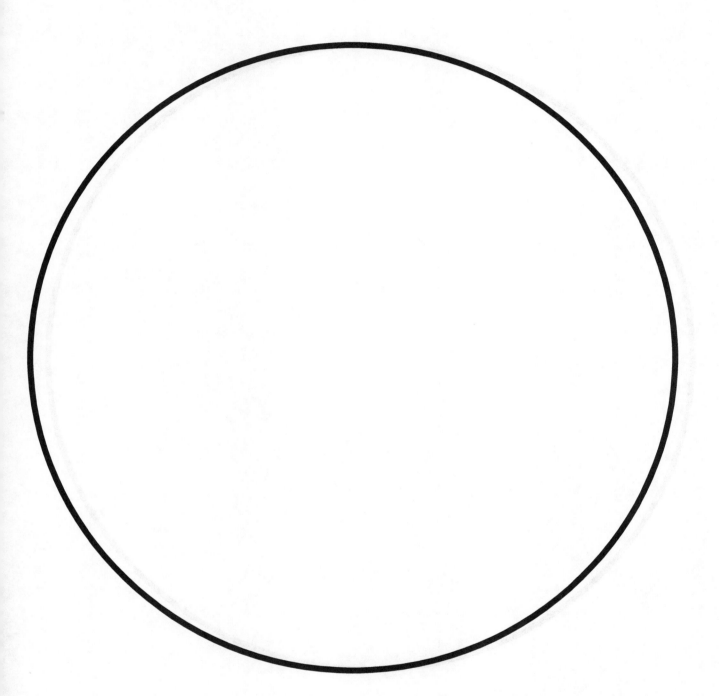

What is under my microscope?

_____

Date _____

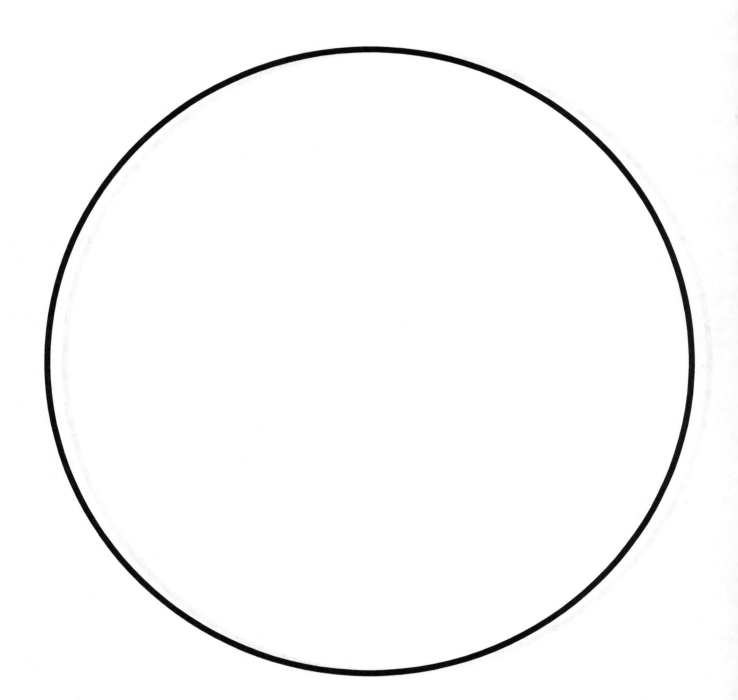

What is under my microscope?

_____

Date _____

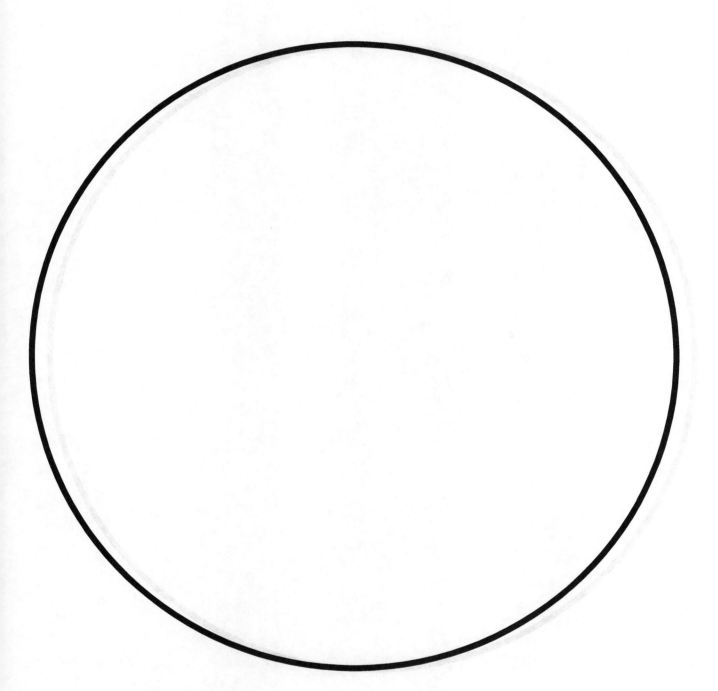

What is under my microscope?

_____

Date _____

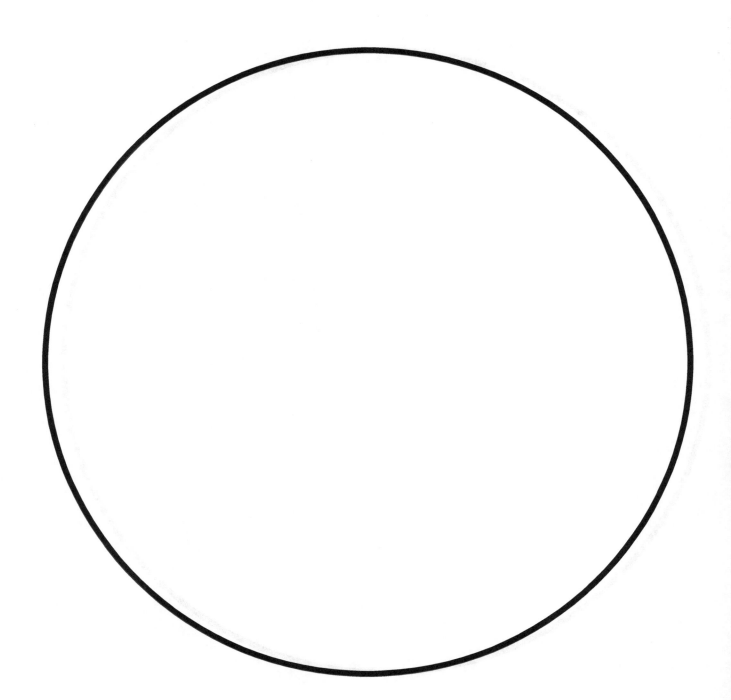

What is under my microscope?

_____

Date _____

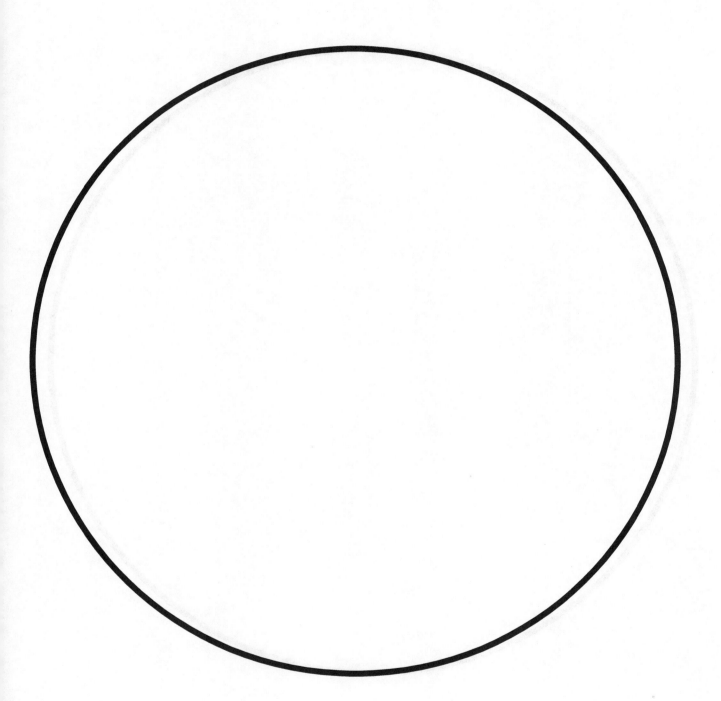

What is under my microscope?

_____

Date _____

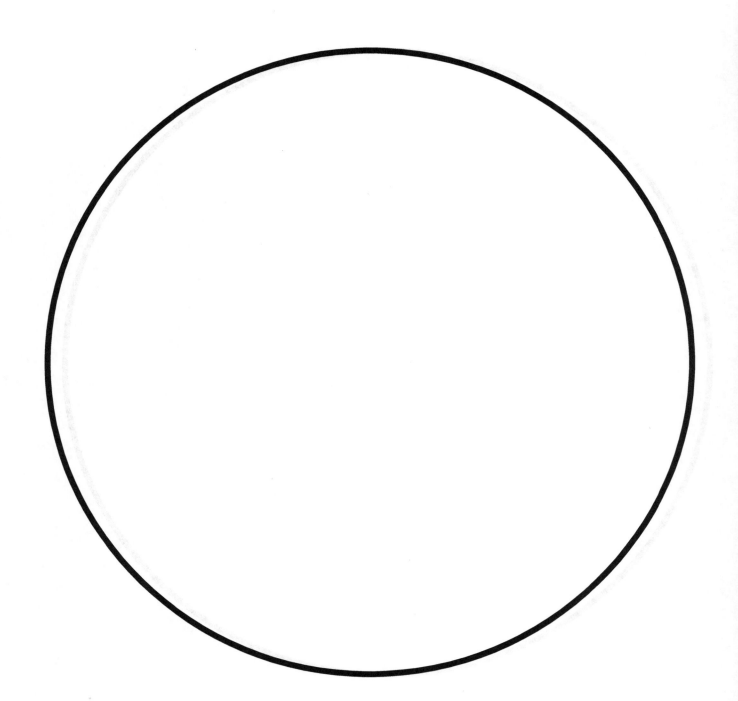

What is under my microscope?

_____

Date _____

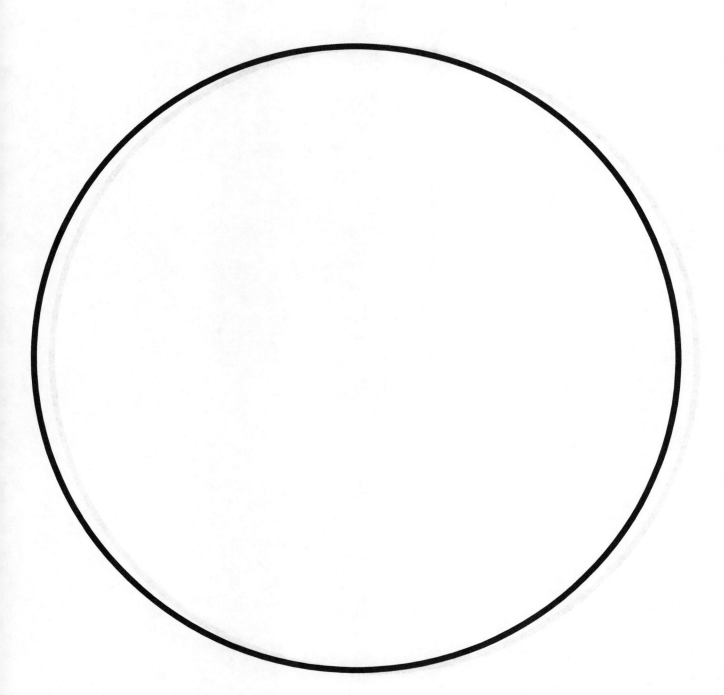

What is under my microscope?

_____

Date _____

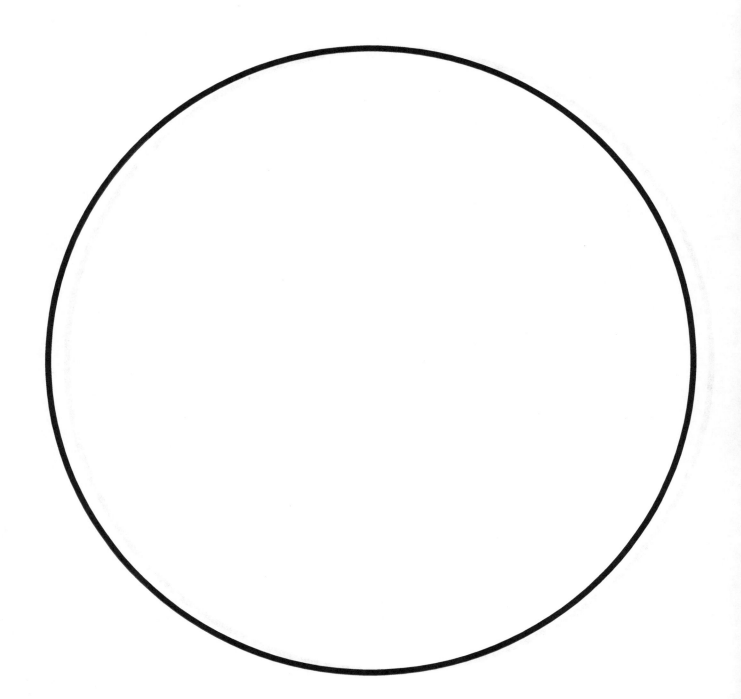

What is under my microscope?

_____

Date _____

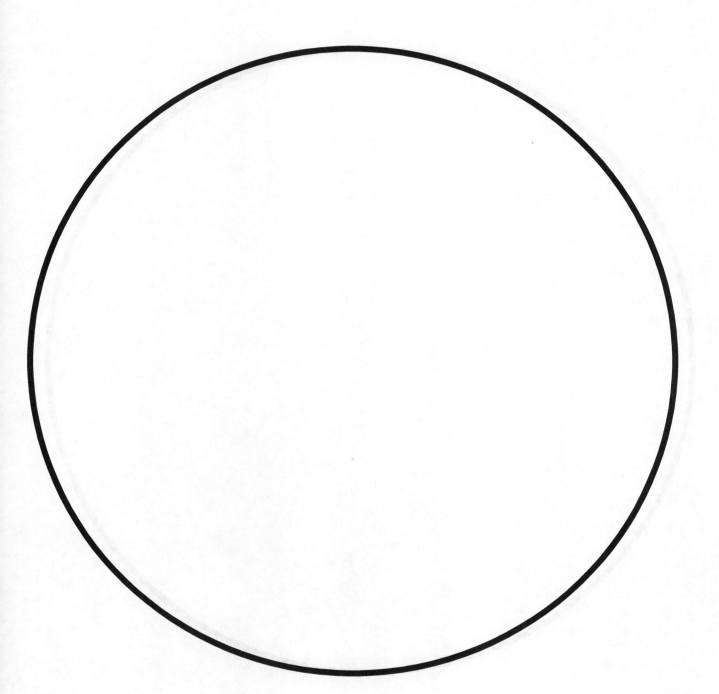

What is under my microscope?

_____

Date _____

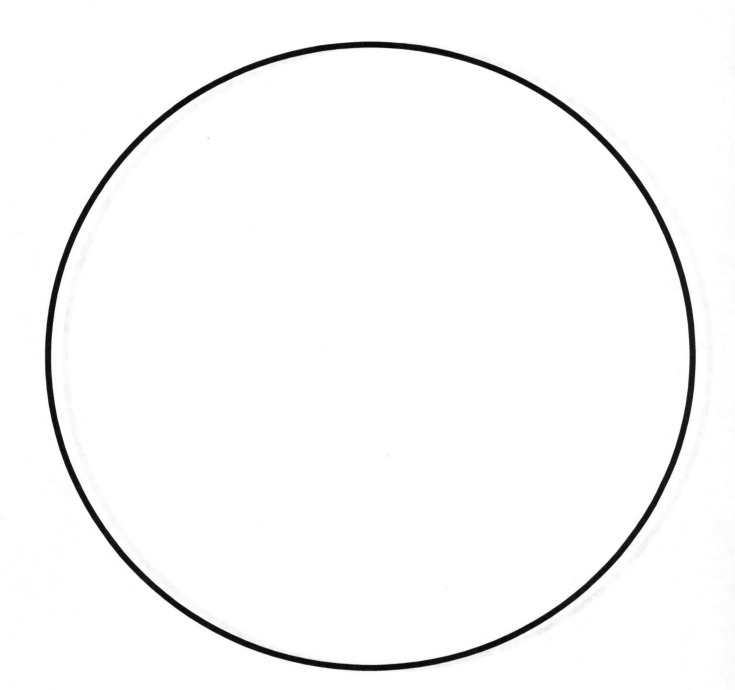

What is under my microscope?

_____

Date _____

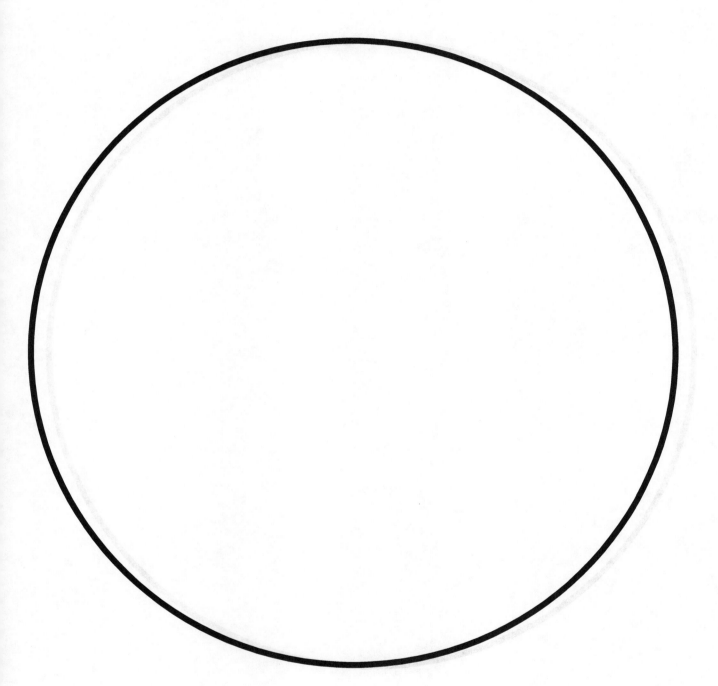

What is under my microscope?

_____

Date _____

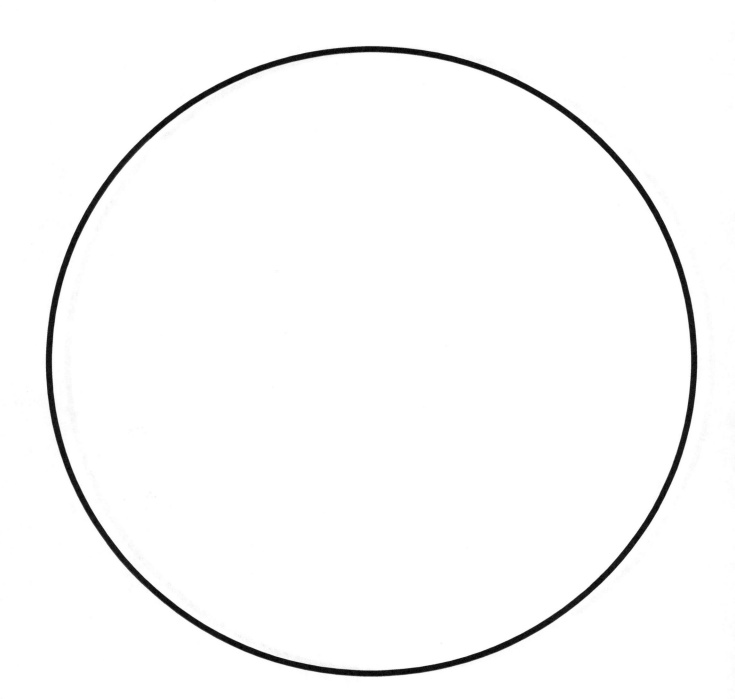

What is under my microscope?

_____

Date _____

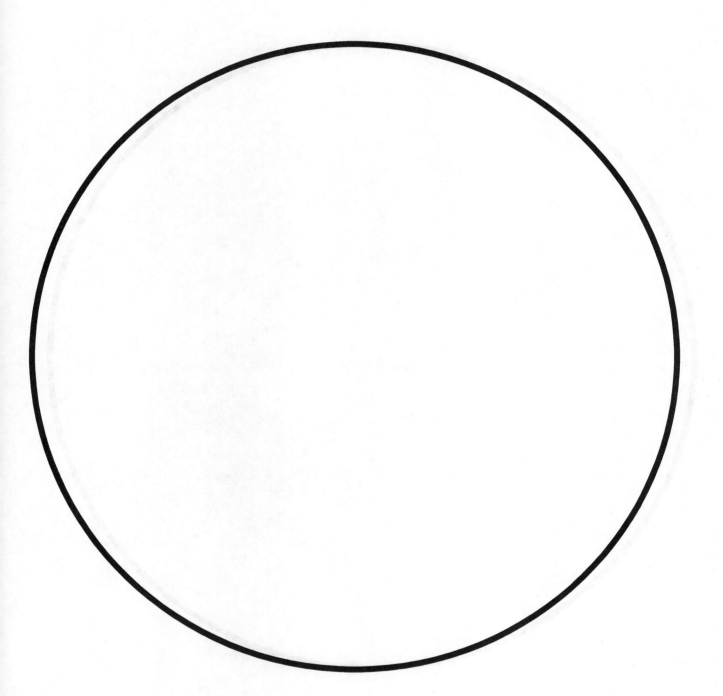

What is under my microscope?

_____

Date _____

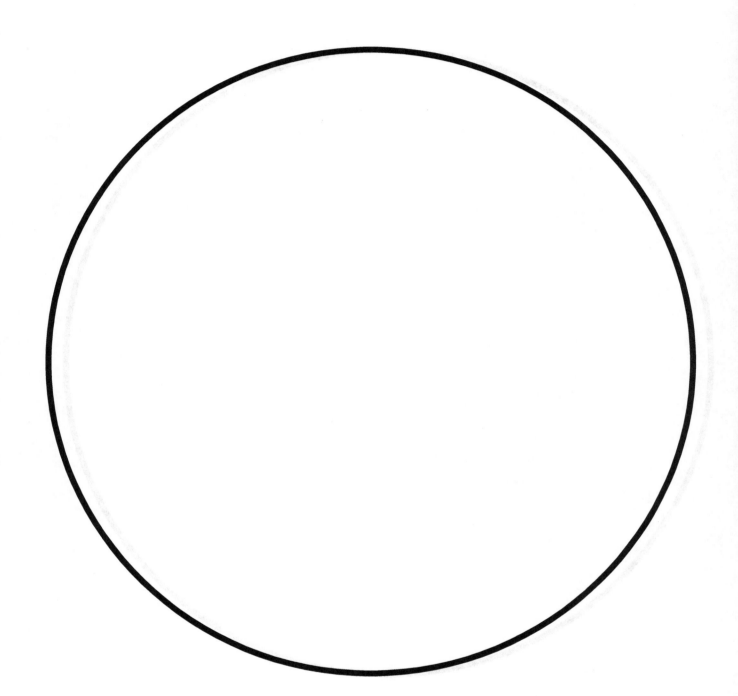

What is under my microscope?

_____

Date _____

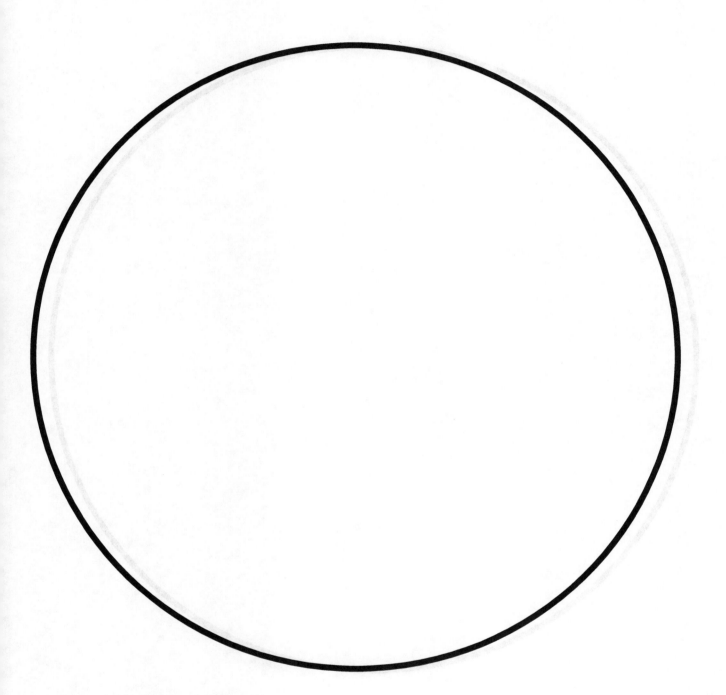

What is under my microscope?

_____

Date _____

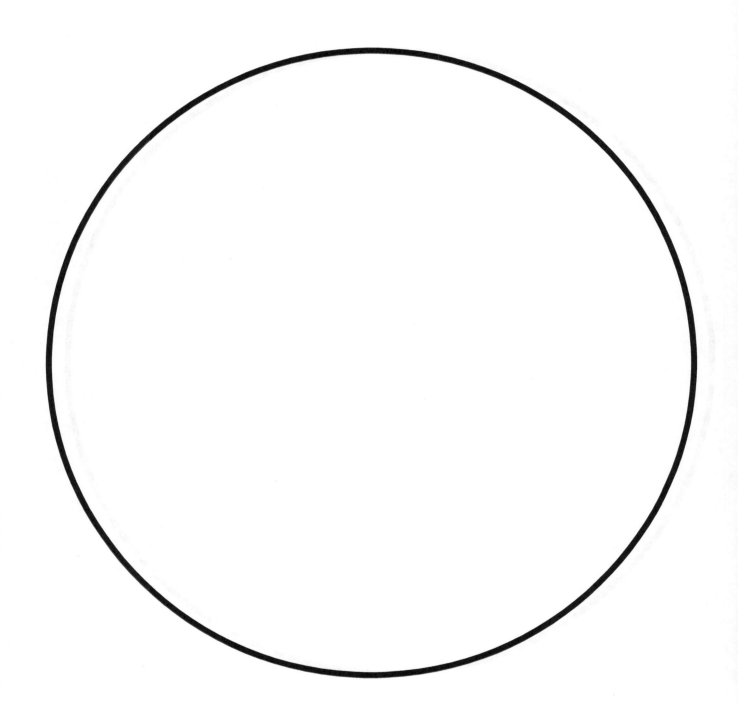

What is under my microscope?

_____

Date _____

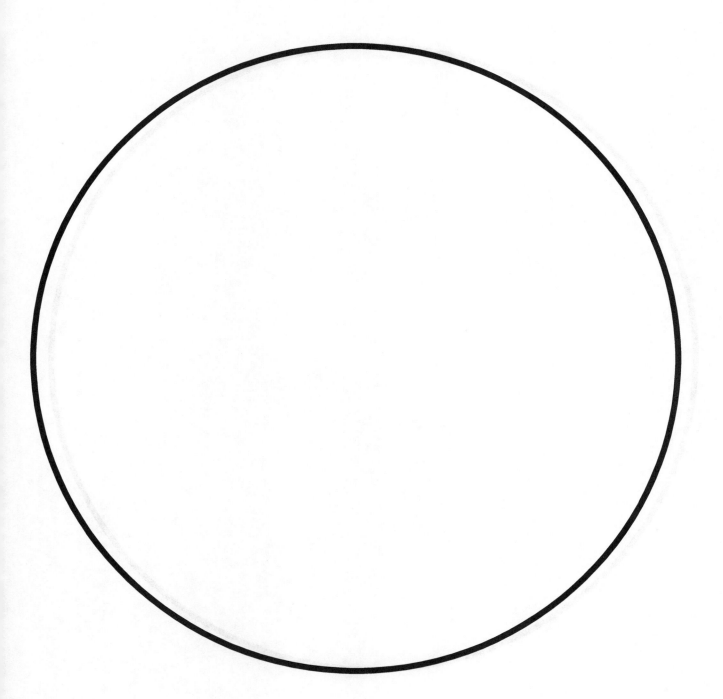

What is under my microscope?

_____

Date _____

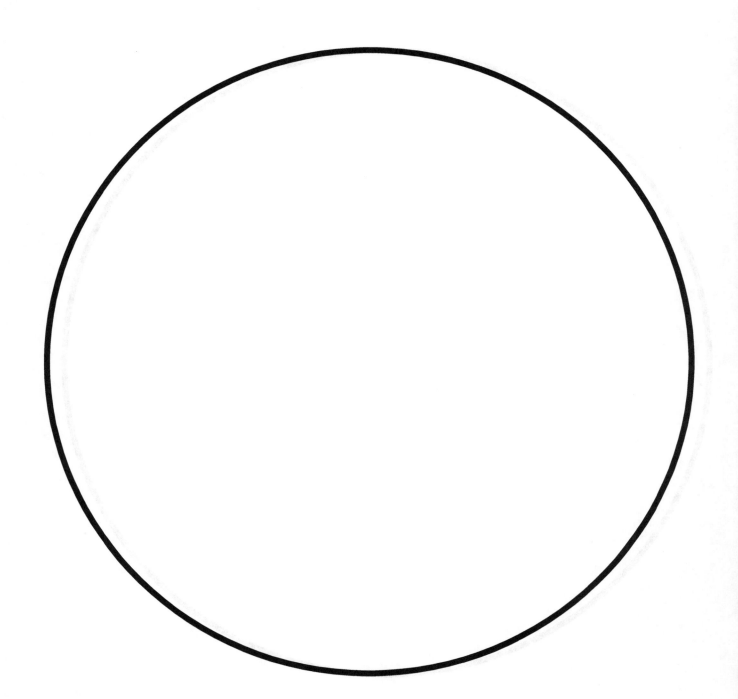

What is under my microscope?

_____

Date _____

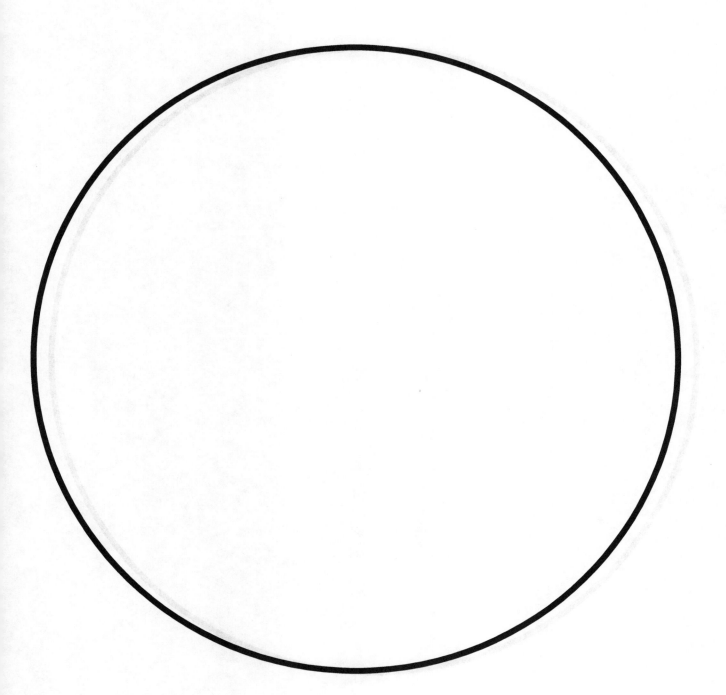

What is under my microscope?

_____

Date _____

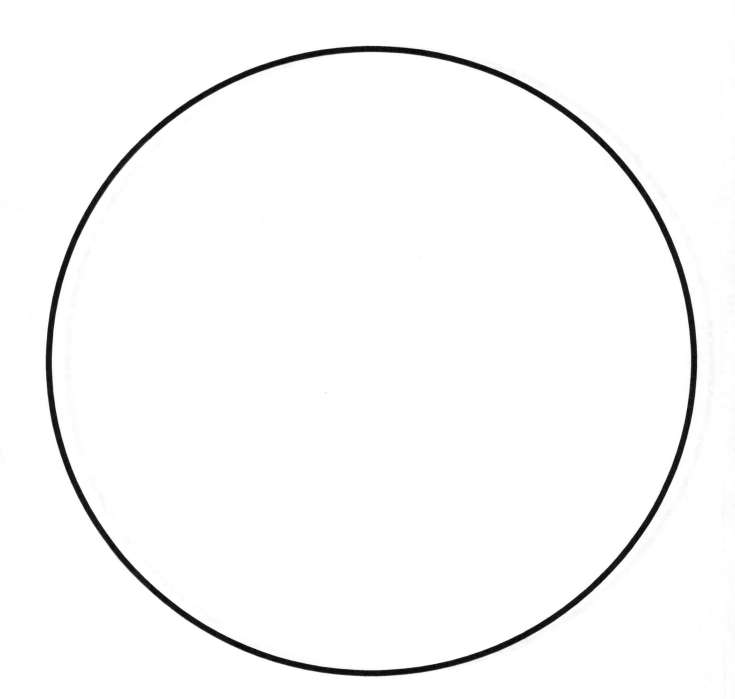

What is under my microscope?

_____

Date _____

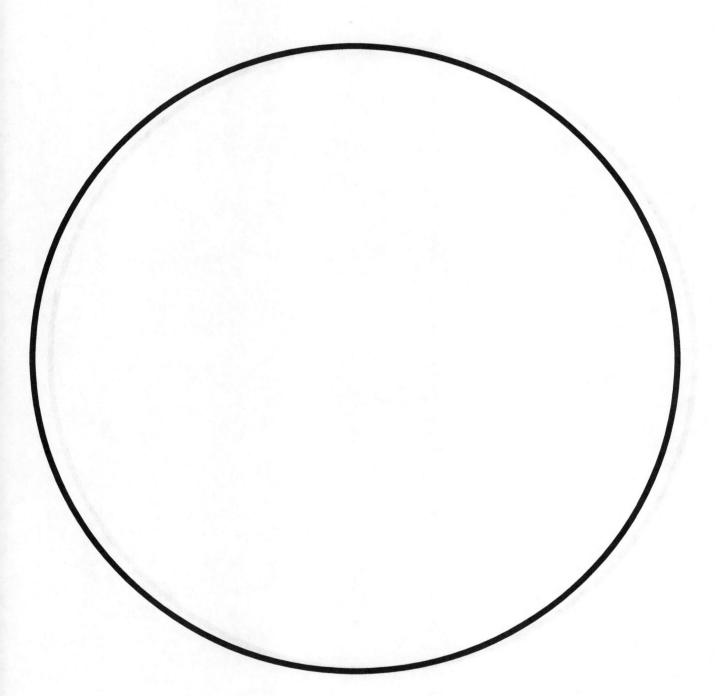

What is under my microscope?

_____

Date _____

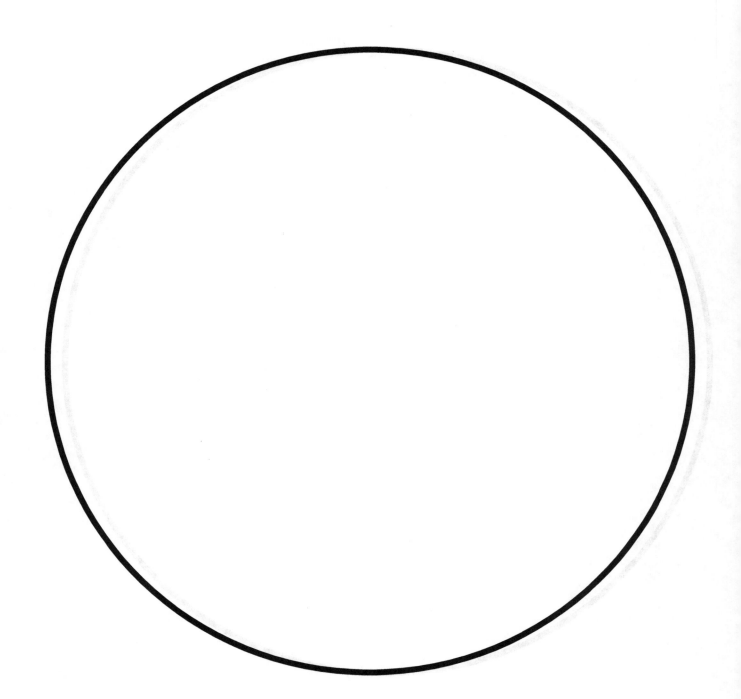

What is under my microscope?

_____

Date _____

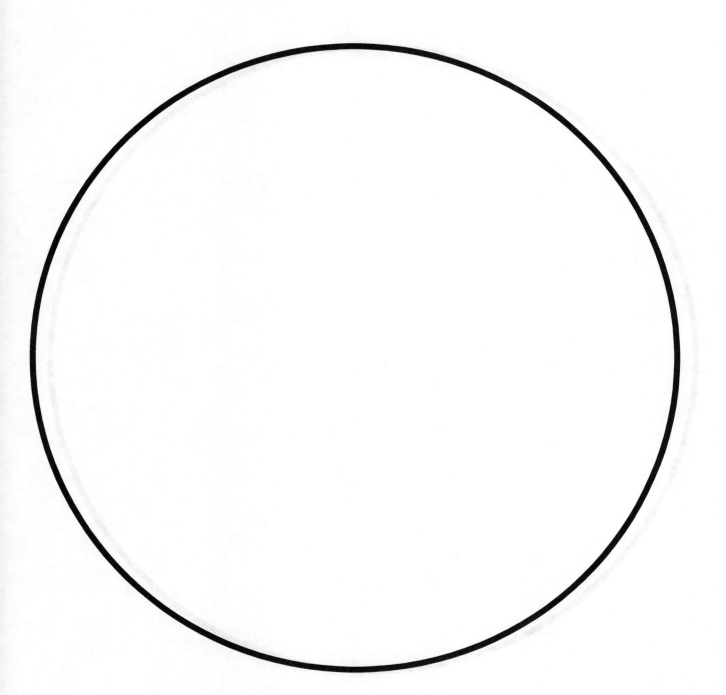

What is under my microscope?

_____

Date _____

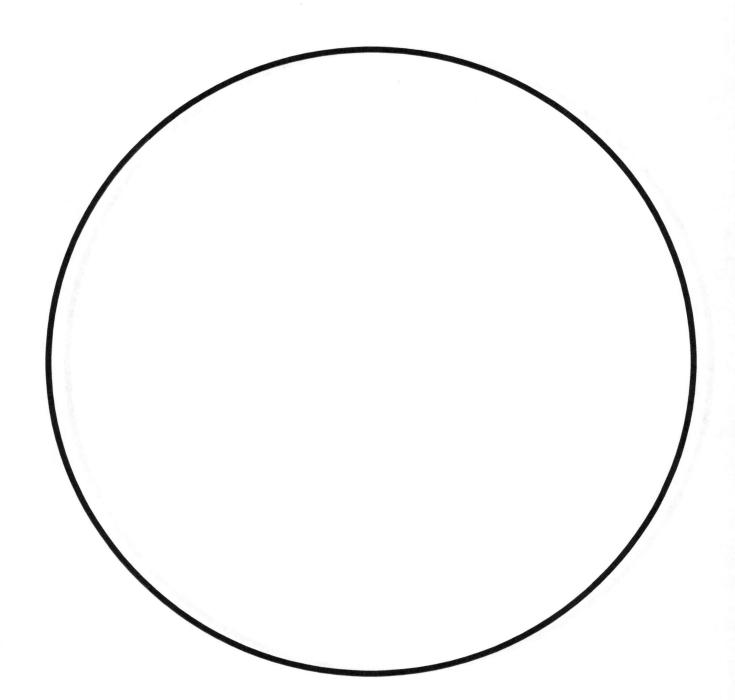

What is under my microscope?

_____

Date _____

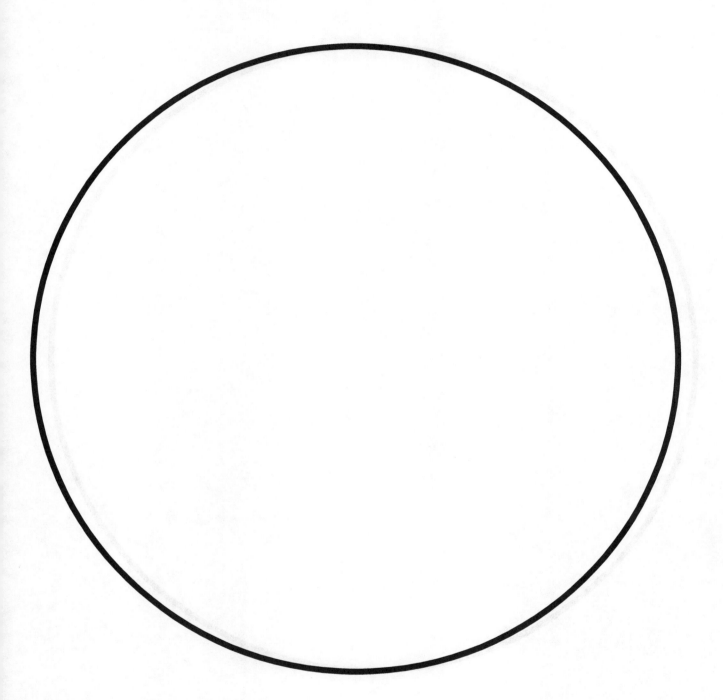

What is under my microscope?

_____

Date _____

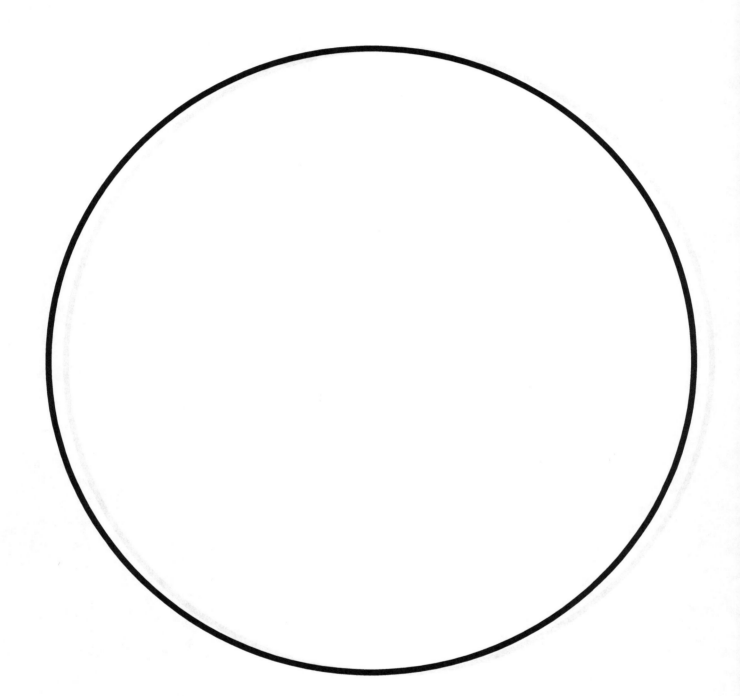

What is under my microscope?

_____

Date _____

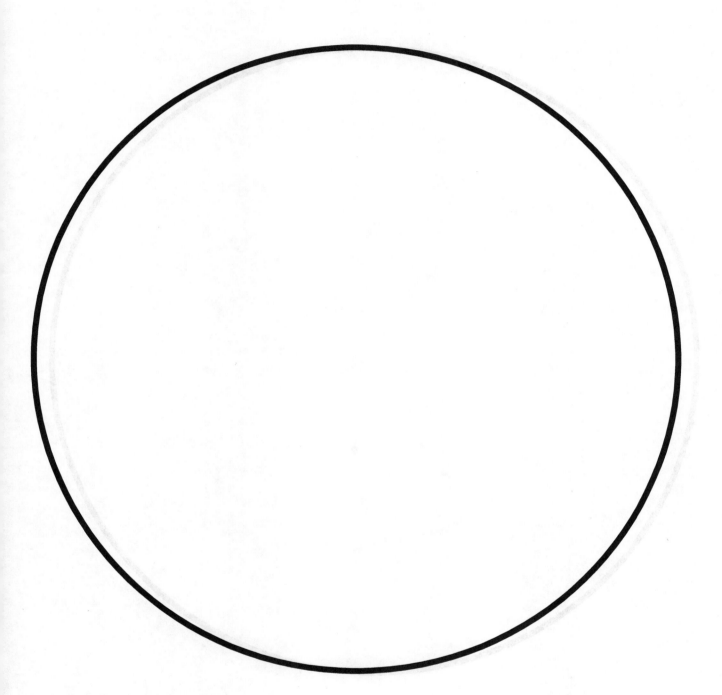

What is under my microscope?

_____

Date _____

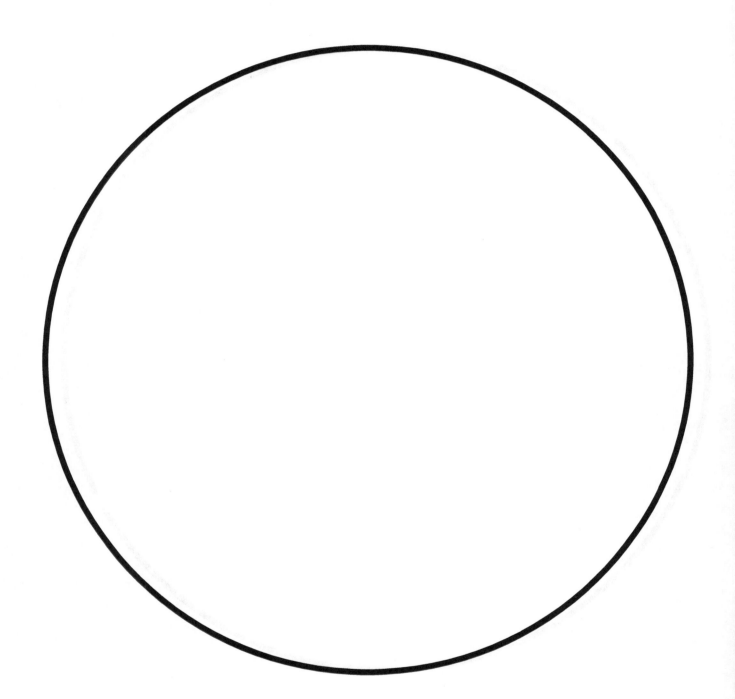

What is under my microscope?

_____

Date _____

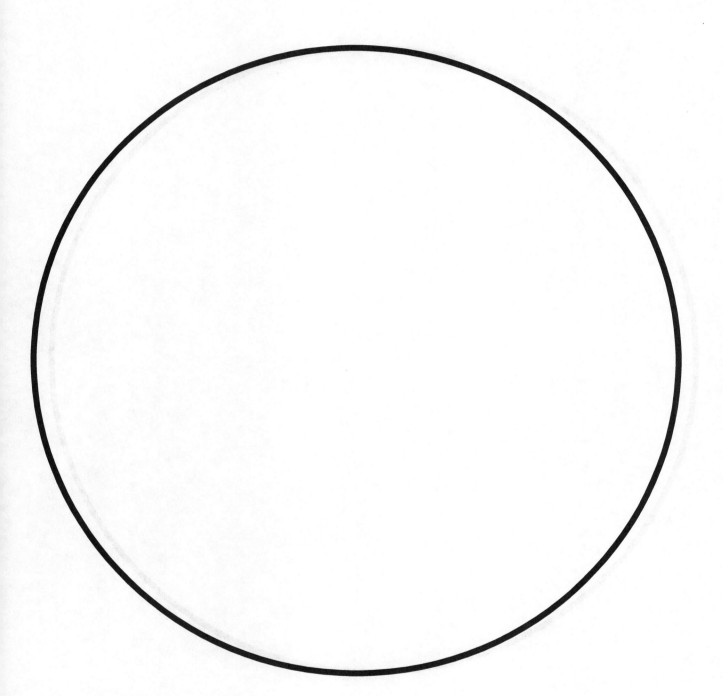

What is under my microscope?

_____

Date _____

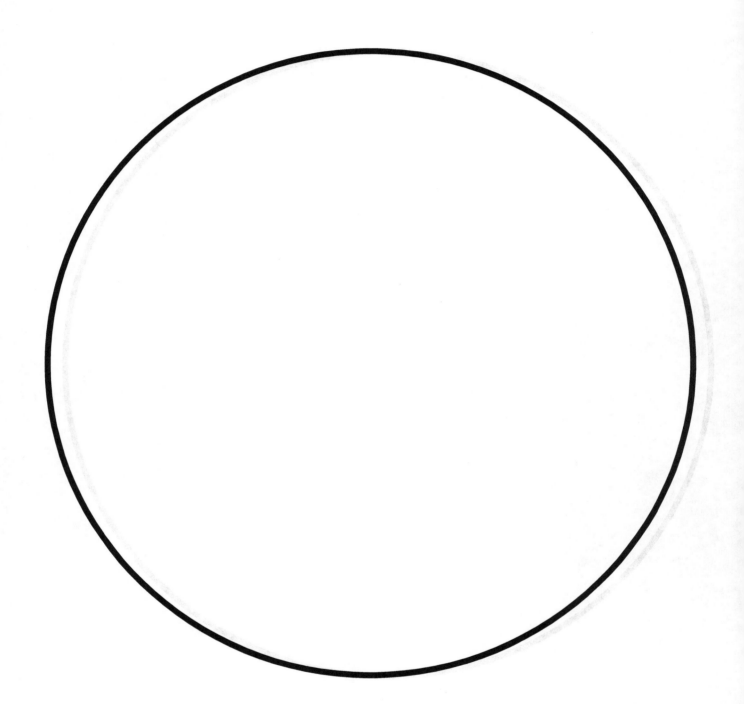

What is under my microscope?

_____

Date _____

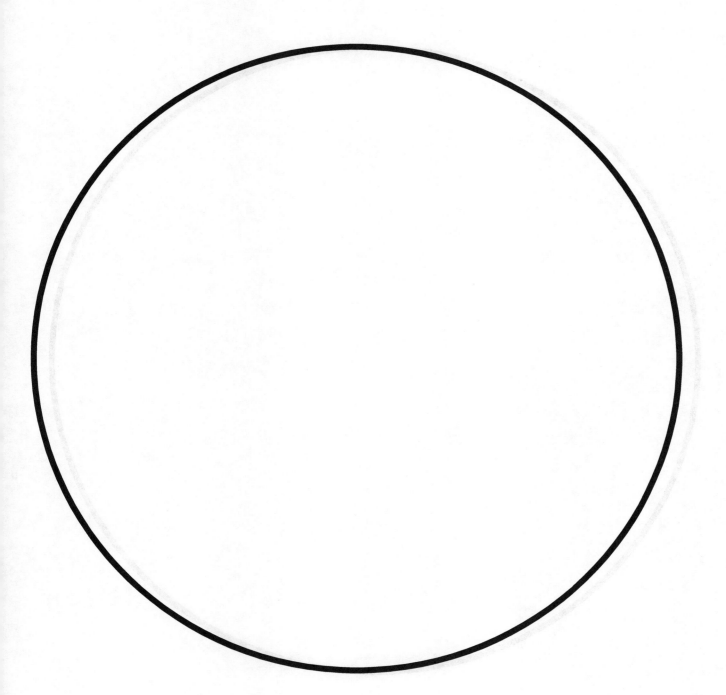

What is under my microscope?

_____

Date _____

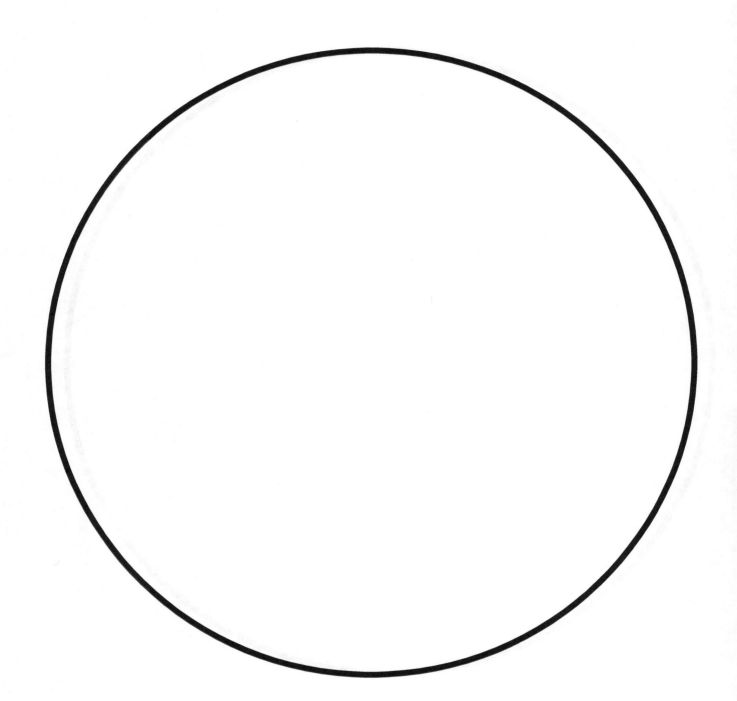

What is under my microscope?

_____

Made in United States
Troutdale, OR
11/28/2024

25431370R00058